Dawn of the Horse Warriors

IN MEMORIAM

Professor Seton Lloyd
(1902–1996)

My teacher, who gave me a chance when
I needed someone who believed in me

Dawn of the Horse Warriors

Chariot and Cavalry Warfare, 3000–600 BC

Duncan Noble

Pen & Sword
MILITARY

First published in Great Britain in 2015 by
Pen & Sword Military
an imprint of
Pen & Sword Books Ltd
47 Church Street
Barnsley
South Yorkshire
S70 2AS

ISBN 978 1 78346 275 9

A CIP catalogue record for this book is available from the British Library

Typeset in Ehrhardt by
Mac Style Ltd, Bridlington, East Yorkshire
Printed and bound in the UK by CPI Group (UK) Ltd,
Croydon, CRO 4YY

Pen & Sword Books Ltd incorporates the imprints of Pen & Sword Archaeology, Atlas, Aviation, Battleground, Discovery, Family History, History, Maritime, Military, Naval, Politics, Railways, Select, Transport, True Crime, and Fiction, Frontline Books, Leo Cooper, Praetorian Press, Seaforth Publishing and Wharncliffe.

For a complete list of Pen & Sword titles please contact
PEN & SWORD BOOKS LIMITED
47 Church Street, Barnsley, South Yorkshire, S70 2AS, England
E-mail: enquiries@pen-and-sword.co.uk
Website: www.pen-and-sword.co.uk

Contents

List of Illustrations

Acknowledgements

A book like this is the result of the contributions of many people. To them, many of whom are alas no longer with us, whether mentioned by name or not, goes my deep gratitude.

They include, among my academic teachers and colleagues, Professor Seton Lloyd, Professor D. J. Wiseman, Professor Stuart Piggott, and Mary Aitken Littauer, For great support during my efforts to investigate the Sumerian battle wagon there was, at the BBC, the late Paul Johnstone. And for his trust in me as a writer, Philip Sidnell of Pen and Sword. Thanks are due to Alan Duncan who drew the maps and to the staff of the British Museum and the Louvre who provided me with photographs of objects from their collections as illustrations

For looking after my horses and introducing me to the complexities of dressage and equine schooling, Sue Wheeler-Adams of the Ox House Riding School and Becky Miles of the Bryngwyn Riding Centre.

A debt of gratitude goes to my partner, Vicky Bernays, for her sterling work and Herculean efforts in editing and proofreading my text on a subject that was new to her.

In gratitude for the companionship and pleasure I have received from our association I must mention at least the most memorable ones of the hundred or so horses I have known and ridden or driven. In approximate chronological order they are Polly, the Clydesdale draught horse who at an early age awakened my passion for horses, Thruster and Charlie on whose backs I learned to ride, Chimo and Lady and latterly Venture West, Percy, and Otto who spent hours walking in circles when they would much rather have been standing in a field and eating grass.

Finally, gratitude must be expressed to four donkeys who had my life in their hands, or hooves, Dougal, Cinnamon, Alfie and Bruno, who endured a noisy battle wagon rumbling at their heels.

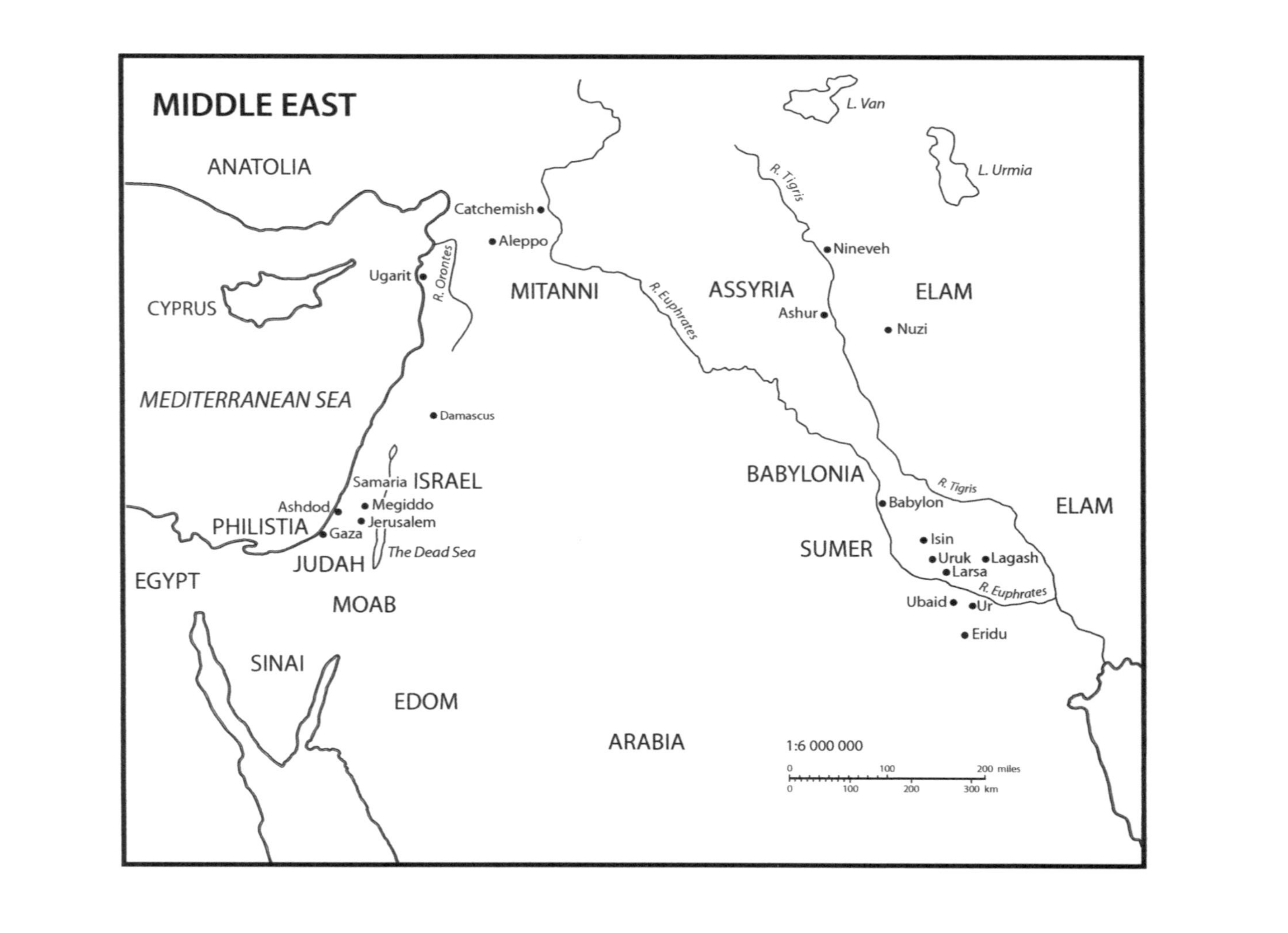
MIDDLE EAST
ANATOLIA
CYPRUS
MEDITERRANEAN SEA
Catchemish
Aleppo
Ugarit
R. Orontes
MITANNI
Damascus
Samaria
ISRAEL
Ashdod
Megiddo
Jerusalem
PHILISTIA
Gaza
The Dead Sea
JUDAH
EGYPT
MOAB
SINAI
EDOM
ARABIA
R. Euphrates
ASSYRIA
Ashur
R. Tigris
Nineveh
L. Van
L. Urmia
ELAM
Nuzi
BABYLONIA
R. Tigris
Babylon
ELAM
SUMER
Isin
Uruk
Lagash
Larsa
R. Euphrates
Ubaid
Ur
Eridu
1:6 000 000
0
100
200 miles
0
100
200
300 km

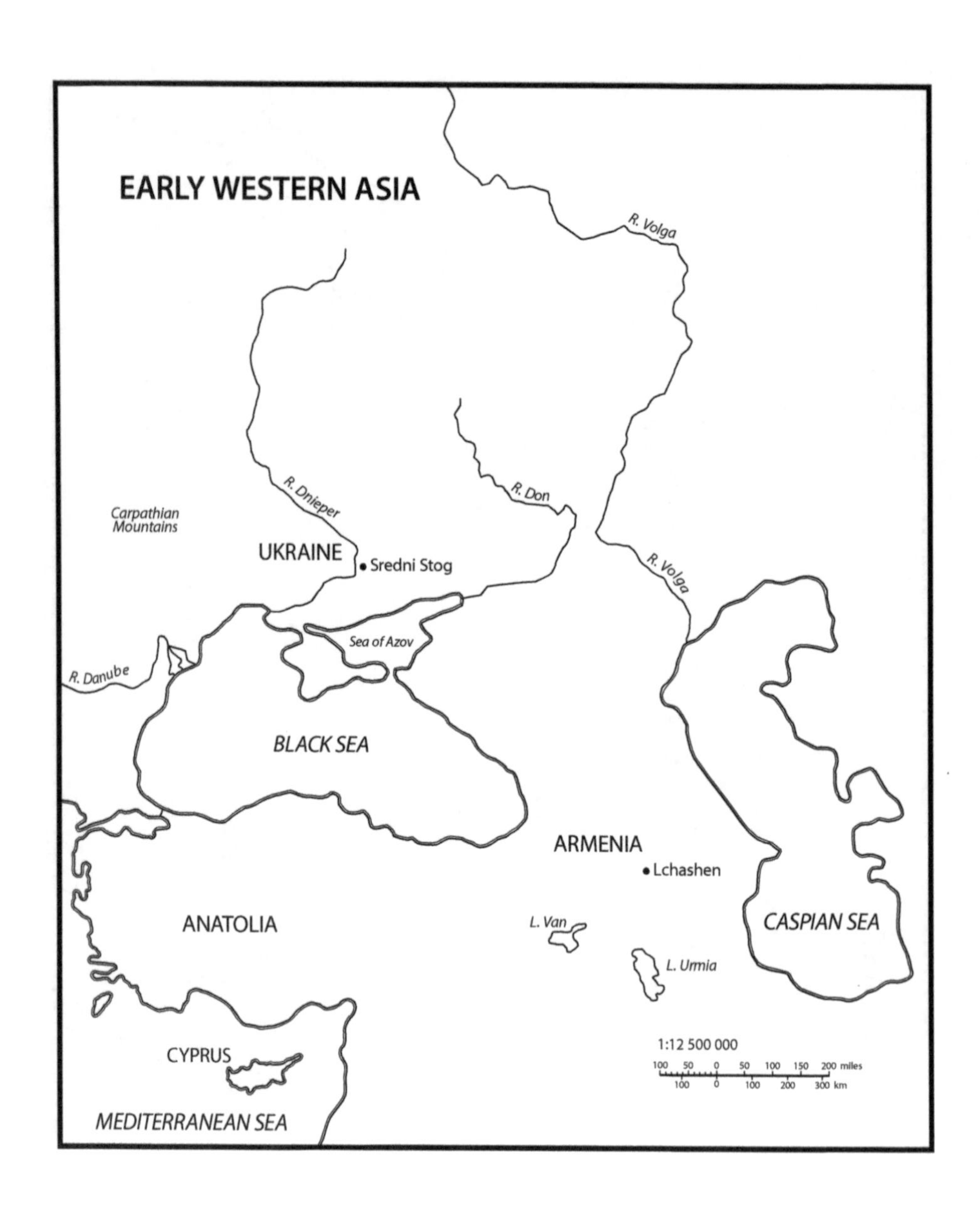
EARLY WESTERN ASIA
R. Volga
R. Dnieper
R. Don
Carpathian Mountains
UKRAINE
Sredni Stog
R. Volga
Sea of Azov
R. Danube
BLACK SEA
ARMENIA
Lchashen
ANATOLIA
L. Van
L. Urmia
CASPIAN SEA
1:12 500 000
100 50 0 50 100 150 200 miles
100 0 100 200 300 km
CYPRUS
MEDITERRANEAN SEA

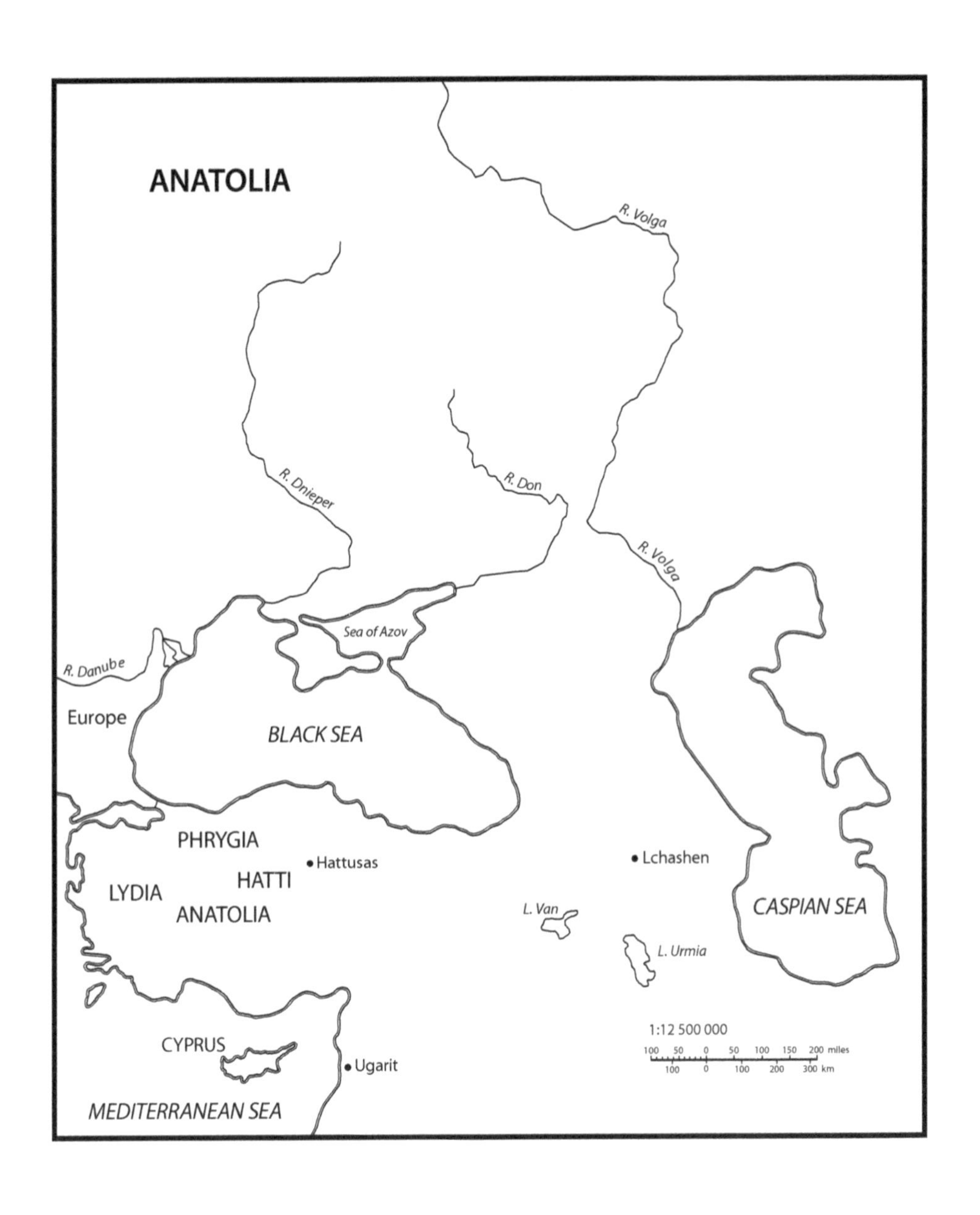
ANATOLIA
R. Volga
R. Dnieper
R. Don
R. Volga
Sea of Azov
R. Danube
Europe
BLACK SEA
PHRYGIA
Hattusas
HATTI
LYDIA
ANATOLIA
Lchashen
L. Van
L. Urmia
CASPIAN SEA
1:12 500 000
100 50 0 50 100 150 200 miles
100 0 100 200 300 km
CYPRUS
Ugarit
MEDITERRANEAN SEA

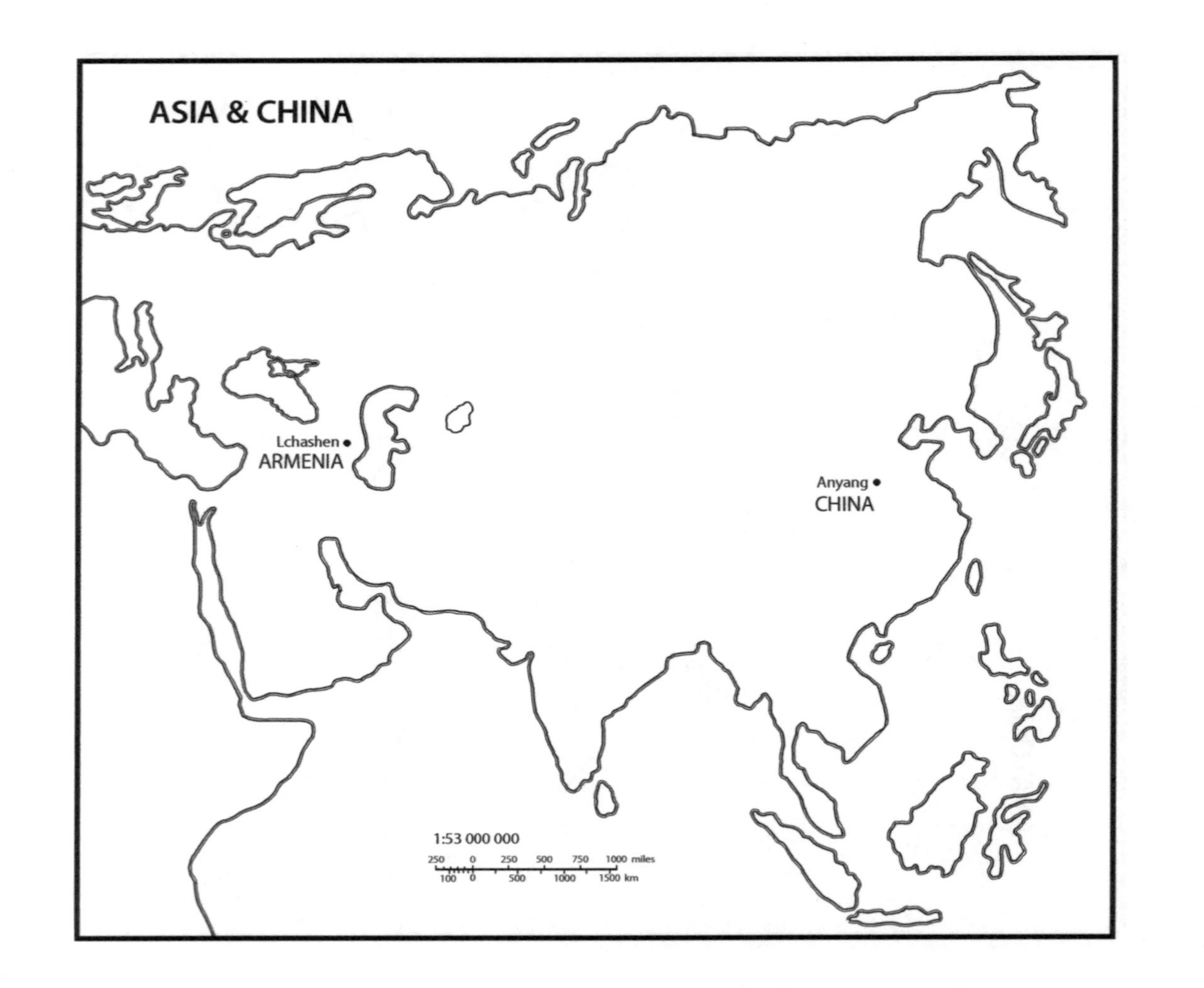
ASIA & CHINA
Lchashen
ARMENIA
Anyang
CHINA
1:53 000 000
250 0 250 500 750 1000 miles
100 0 500 1000 1500 km

Introduction

For a thousand years of the Western Asiatic Bronze Age from around 1400 BC to 400 BC the horse-drawn chariot was the supreme prestigious fighting vehicle. It was only from about the 700s of the 1st millennium BC that starting with the Assyrians it began to be replaced by cavalry.

Its employment spread in the Bronze Age in the 2nd millennium BC beyond Western Asia westwards to Mycenaean Greece and eastwards to China. At that time every country in Western Asia that aspired to the status of a major power had chariots, in some cases thousands of them.

This is the story of how chariots dominated the military inventories of states between Greece and China for a millennium. Then, when the technology of riding horses had improved, chariots gave way to cavalry; men fighting while sitting on the backs of horses instead of standing or crouching on wheeled vehicles being pulled by the animals.

This book tells the tale of that phase in human technological development, when the early wars of expansion in Western Asia went from being slow slogging matches between infantry to become wars of movement, speeding battle up from five to twenty miles an hour.

This is a study of the history of the invention and adoption of new methods of undertaking fast mobile warfare, first with the battle wagon, then the chariot, and finally cavalry, by the major military powers, principally in the Middle East, in the land that lay between Greece and China. Chariots were the new fast fighting vehicles. They began to appear in the period between the rise of the Sumerian city states around 3000 BC and the fall of the Assyrian Empire in 612 BC.

As we are dealing substantially with chariots, we should start by defining what we mean when we talk about that major weapon that filled the military inventories of countries that aspired to dominance. And it cost their treasuries dearly for a thousand years.

A chariot was a light two-wheeled military vehicle pulled by two or four horses harnessed abreast. It carried a driver and between one and three fighting men, armed principally with bows and arrows and in some instances with stabbing spears or throwing javelins. They were exciting, prestigious, and expensive to procure and maintain. In the 2nd and 1st millennia BC even moderate sized military powers possessed hundreds of them. The Great Powers of the ancient world, principally in the Middle East, maintained and could field thousands of them in battle. If the chariot was occasionally used as a ceremonial vehicle in which a king could travel in a procession, or as a sporting vehicle from which he could display his manly hunting prowess in killing dangerous animals like wild bulls or lions, it was still essentially a military fighting vehicle.

There were four wheeled wagons that served as agricultural vehicles as well as two-wheeled carts used for similar purposes. At religious festivals in the towns of Western Asia vehicles resembling chariots were used to carry images of the local god and we have pictures and small clay models of them. But they do not qualify as chariots. So we are not concerned with them in these ecclesiastical functions, even if on occasions they were military vehicles used for religious purposes. We are dealing with military fighting vehicles being used in warfare.

Three preconditions for the development in the 2nd millennium BC of the light fast chariot have been postulated by Piggott (1983) that suggest that it originated in Western Asia.

First, there should be a social structure in which a king was intent on demonstrating his importance and where his ongoing prestige was important. His country should be following an expansionist policy as a rival to its neighbours. These should live in open and reasonably flat country and be less militarily advanced than the expansionist country. Western Asia was ideal.

Next, there should be available a suitable draught animal, the horse, although that animal was also found in Europe.

Finally, technology should have advanced to the extent that a light two-wheeled vehicle with spoked wheels could be maintained, repaired, and kept in service.

Western Asia satisfied all those requirements. Western Europe was too heavily forested while Mycenaean Greece was near enough to the Levant to borrow the idea of the chariot from there. Western Europe did not take up the chariot till well after Western Asia adopted it; not until European chiefdoms became larger and there was a nearby expansionist power that had chariots. That was Rome.

Chariots cannot be parcelled neatly off into national types, with the possible exception of those of the Egyptians and the Assyrians. The Egyptian and Assyrian chariots are the two distinct types of whose appearance we have good evidence. The Egyptians went in for very light chariots with a crew of one or two men, while the Assyrians preferred heavy ones with a complement of up to four men aboard. Both of those powers probably supplied chariots to countries whose policies were aligned with their own and which were useful to them. It is conceivable that countries that fell within the spheres of influence of these two great powers could have incorporated details of their military suppliers' chariots in any warlike vehicles of their own that they might have built. But any knowledge we have of the design of chariots of Middle Eastern countries other than those of Egypt and Assyria has to be based on pictures of the armies of those countries where they appear in Egyptian or Assyrian reliefs or wall paintings. So often we do not know in any detail what the chariots of the lesser powers looked like.

This study inevitably concentrates on war. War is as old as civilization. What we mean by that statement depends on which definition of the words 'War' and 'Civilization' we prefer.

The Prussian philosophical theoretician on war, Major General Carl von Clausewitz, writing in his book On War (*Vom Krieg*) between 1815 and 1830 and deriving his ideas from his war experience in the Napoleonic Wars of 1792–1814 defined war as follows, 'War is simply the continuation of political intercourse with the addition of other means.' These means are of course violent ones. Records of human violent strife go back to the Palaeolithic 50,000 years ago, where cave paintings depict bands of men shooting arrows at each other. But these are fights between groups of hunters using their arrows on each other, probably to defend their hunting area from chance intruders. They are not full-time servants of a state.

Here I am defining war as organized violence by a state against another state for political ends, undertaken by men who are, if only for the time being, involved exclusively in the fighting. A general unorganized attack by men of a society, such as mounted nomad horse archers, on the men of another similar group does not count as a war in this definition. The reasons for the war can range from the maintenance of the state's prestige with other states to the acquisition and retention of wealth and political power or the control of natural resources or trade. It would not be profitable here to involve ourselves in the legalistic argument whether an attack by a state on a section of another state with that other state's permission constitutes a war.

Dates are given in the well understood form of BC (Before Christ) after the numeral instead of the fashionable American form BCE (Before Common Era) and where appropriate as CE (Common Era), instead of the now less fashionable AD Anno Domini, which is awkward to use in conjunction with the other as it has to come before the numeral.

I should pause here to explain the form in which ethnicity is expressed by historians of the ancient Middle East. Apart from Egypt and Turkey, the present political states of the Middle East are modern twentieth-century creations. With the exception of Israel that dates only from 1948 CE, they were imposed on the inhabitants of the Ottoman Empire when it was dismantled by Britain and France after the First World War. In the Bronze Age of the second millennium BC it was divided up into states that no longer exist and their inhabitants were known by names such as Philistines and Canaanites that are no longer used. So historians identify the origins and relationships of ancient peoples and states by their languages.

The Assyrians and their southern neighbours the Babylonians spoke a language that belonged to the still existing family of Semitic languages. There are three Semitic languages still being spoken. They are Arabic, Hebrew, and Aramaic (the language of Christ). So both the Palestinian Arabs and Israelis are Semites, and the Arabs strongly disapprove when called anti-Semitic, since it means that they hate themselves. In antiquity there were numerous Semitic languages spoken, which is hardly surprising for an area where to this day the local dialect changes every two kilometres. A language has been defined as a dialect that has an army.

We do not know where the Assyrians and Babylonians came from, although we think it was possibly Arabia. The stages by which they occupied what became Assyria and Babylonia are not recorded by history. Little is known about the military and political history of Babylonia, except where it impinged on Assyria. But a great deal is known about Assyria before the collapse of its empire in 612 BC. Every year it fought a war in which chariots formed a substantial part of the army.

There are several reasons for conquering a country and the history of the Middle East is one continual round of conquest and subjugation. One reason for waging war is to acquire land in which your own people can settle. Another is to acquire wealth. If it is the land you want, the indigenous population is surplus to requirements and is usually massacred. If it is wealth, then you want the local people to remain and work the land for your benefit in the form of tribute instead of paying taxes to a government chosen by themselves. And a war may be defensive, to prevent any of those fates overtaking your country. All too often a war that was originally defensive soon takes the form of an aggressive one.

Here I want for convenience and clarity to make a distinction that I shall come back to later between the terms domestication and taming of animals as I use them.

Domestication, by my way of it, is the reduction of an animal to a state where it will live alongside humans and can be used as a source of meat, milk, hides and leather. Taming is the process whereby that animal subjects itself to work for humans by performing acts that are useful to humans such as, in the case of the horse, drawing a wheeled vehicle or allowing a human to sit on its back and direct its direction and the pace at which it moves. By that definition a dog is domesticated and tamed. A cat is domesticated but not tamed.

M. A. Littauer in her 1977 study makes a most useful distinction between carts and chariots. She says that the former with solid or spoked wheels were confined to the transport of men or goods for peaceful purposes and when they had two wheels the axle was beneath the centre of the body. When they were carrying men the passengers sat in the vehicles.

Chariots, by her definition, had spoked wheels and with certain notable exceptions had the axle at the rear of the body. Except in China the occupants stood in them and the vehicles were used for military, sporting or ceremonial purposes.

We must exempt the four wheeled Sumerian battle wagon from Littauer's definition. It was definitely a military vehicle and neither a wagon nor a chariot.

And the terms chariotry and cavalry also demand definition. A handful of arrow-shooting nomads riding in vehicles or sitting on their horses do not constitute a chariotry force or a cavalry unit as I would define those. Chariotry and cavalry as I use the words are organized and disciplined units of an army working in planned cooperation.

One of the problems of archaeological research into ancient chariots and cavalry is that it is a highly specialized branch of historical research, which demands a good knowledge of horses and vehicles, a knowledge that has to be gained through practical experience with horses and horse-drawn vehicles. That is not information that is part of the fund of knowledge of the usual historian or archaeologist.

Before the invention of printing with moveable type in 1450 CE, many aspects of human activity were undertaken by illiterate men who taught by example and could not write text books on their subjects. Among these were horse riding, driving, and weapon handling instructors, and many others whose manual skills did not leave physical remains or texts that can be excavated.

Writing in the ancient world was a highly specialized skill that was often hereditary and was in the hands of professional scribes, religious teachers and priests and civil servants. What they wrote were generally the minutiae of administration, which had to be recorded, accounts justifying and glorifying the political and military achievements of their masters, legal matters and religious exhortations. They, and the people who ordered the production of their texts, were not interested in the details of technical processes and skills such as the training of horses or the technical details of military tactics and technology.

Here it is worth looking at how archaeological research and excavation is organized. An archaeologist may discover traces of a site belonging to a culture in which he is interested. Alternatively he may be offered the chance to excavate, or be ordered to excavate a site by his employers, because it is threatened with demolition in the process of building redevelopment or of flooding by a proposed dam. The last was often the case in the former Soviet Union where the archaeology was of a high standard but the time available for excavation and subsequent publication was limited. If the archaeologist has himself found the site he wants to excavate from the surface scatter of broken pottery and he has done the necessary reading up of the subject, he then has to find finance. And excavation is an expensive business. A lot of people have to be transported, housed and fed, and if they are specialists such as conservators or epigraphists, the salaries they would have received from the museums or other institutions who normally employ them have to be paid.

This money has to come from institutions such as learned societies, archaeological charities, museums, universities, who all want a return on their money. Learned societies and the like want a prestigious publication in which their name will be prominently displayed. Museums want objects that will enhance their collections. So excavation has to produce a grand palace, temple, or other building, preferably of a society that was of historical importance at that time and place and if possible is full of interesting finds.

A midden that produced only the bones of a few horses is unlikely to find much financial sponsorship, even if horses have never before been found from that historical period before in that area. And it is only when digging has been going on for some time, at considerable expense, that the director of the excavation finds out what he has really got. It may be an exciting new discovery or else a repetitive collection of no new scientific interest.

Military chariots that can be driven at speed in combat situations are the result of long years of experimentation and improvement in technology and horse breeding and training. The possession of a chariot would be insufficient to provide a useable military vehicle without a long gathered experience of the psychology and physical attributes of the horse.

A short explanation should be made here of the difference between the geographical terms Near East and Middle East. In British English before 1939 Palestine and Turkey were in the Near East. Countries east of them as far as Iraq and Iran were in the Middle East.

From 1945 to date British writers have adopted the United States term of the Middle East for what had before that included the Near East and the Middle East. Academics often use the term Western Asia to cover what is more popularly called the Middle East. It is not geographically specific, but everything east of the Bosphorus is in Asia. Egypt is generally included by journalists in the Middle East because of its Arabic culture and political alignment. Technically it is in Africa, but to include it politically with Africa would be confusing.

Organization of people into armies, and an end to the situation where every man reached for his weapon when danger threatened, had to wait for civilization. There are numerous definitions of the term 'civilization' and of 'civilized' going about. Many of them pose as moral ones and brand as uncivilized the people of whom the writer disapproves. Archaeologists use a simple, and I submit a useful, definition that leaves moral questions aside. It is based on the Latin *civis*, a citizen. In other words a society is civilized when its people live together and there is division of labour. No longer is the political leader also the war leader, even if he accompanies the army to war for public relations reasons. Nor is he also a farmer producing food. The potter and the smith are paid by society, through the purchasers of their wares, as are the professional fighters, the soldiers.

It is not easy to detect when this state has been reached. One can be pretty sure when the state in view has reached such a degree of complexity that it has been necessary to produce writing with which to keep records of stores received and expended that it is civilized. Even better indicators are written records detailing the state's activities. This is dependent on the writing being in an imperishable form.

Direct imperishable written evidence is available from Assyria, Egypt, and China because their writing survived. Assyria's lasted because it was in the medium of clay cuneiform tablets, which survived when dry and when baked were virtually indestructible and in Egypt, apart from reliefs on stone, because the climate was so dry that wall paintings and papyrus documents dried out. From China the texts survived written on strips of bamboo. For the lesser countries that left no useful records of their own we depend on the accounts of their enemies if these were Assyria or Egypt. Even the best written texts we have from these two victor countries are accounts glorifying the martial successes of those states, often leaving out the details that would be of most use to the specialist historian. The present day historian has the problem that a future one writing in five thousand years' time would have if the only surviving accounts of the successes and problems of twenty-first-century governments were the speeches made by party leaders at annual conferences.

When it describes in detail the use of chariots and cavalry in antiquity this book covers principally the period between 3000 BC and 612 BC. The earliest date coincides approximately with the appearance of the fully formed Sumerian city states and their battle wagons, forerunners of the chariots in that area. The latter is the accepted date of the fall of the Assyrian Empire, the pre-eminent military power in the Middle East between around 1200 BC and the 7th century. The Assyrians have left us with more detailed technical information on their chariots than have western Asiatic powers other than the Egyptians and it is therefore inevitable that any study of chariots should lean heavily on Assyrian sources. The Assyrians were also the earliest military power to employ cavalry, that is, bodies of uniformly equipped horsemen trained to fight in cooperation with each other and as part of a disciplined army. Bold armed horsemen fighting independently of each other are not considered as cavalry. So this date allows us to bring the story up to the demise of chariotry as the dominant mobile arm in ancient warfare and its replacement by cavalry.

These two dates are convenient ones for any discussion of the use of horse power in warfare. But inevitably they do not coincide with significant events on the warlike history of all the societies across Europe and Asia such as Central Asia and China that are being discussed. The history of the domestication of the horse started in Central Asia long before 3000 BC and the date of 612 BC is not significant in Chinese history. So in places the survey has extended before and after the two time limits laid down.

The survey starts in the Central Asian forest of the taiga and the steppe, that great stretch of grassland that is the natural homeland of the horse. The steppe starts in Hungary and then goes on after a break for the Carpathian Mountains to Mongolia. Then after Central Asia my survey goes south to the southern reaches of the Tigris and Euphrates rivers, to the Sumerian city states with their battle wagons and the expansionist Assyrians with their chariotry and then cavalry in their highly organized army, to civilized areas. For warfare demands civilization. To the historian the term civilization does not have, or should not have, a moral component. In any case, accepted morals change dramatically over time. Historians are concerned with why things happened, not whether they were justified in modern eyes. The philosophers and theologians can be left to consider the morality of past actions.

From the Mesopotamian powers the study goes on to consider the use of chariots further north and west in Anatolia, Egypt, Palestine and Syria.

Mycenaean Greece is the one chariot-owning European power that falls within the self-imposed time scale and further east we examine what is known of chariots in India and China.

The historian is limited by 'The accidents of excavation'. The world is a large place and archaeological excavation is expensive. A site may be excavated to illuminate historical problems that have nothing to do with chariots. The people there may or may not have used chariots and the remains of the vehicles may be in badly deteriorated condition. The people may have been illiterate like the Sintashta culture people of Siberia and have left no explanation of how they used the chariots they buried. Or their written records may not have been on an imperishable medium. Or, like the Indus Valley texts they may be in a script that we are not yet able to read. You never know in detail what you are going to find till you start digging. It may be exciting new information. Or your efforts may produce only repetitive finds that are well known elsewhere and add little or nothing to our knowledge of human history.

The period of 3000 BC to 612 BC was one of vast dynamic changes throughout the civilized world. During its two thousand, three hundred and eighty-eight years it saw the rise and then collapse of new international powers that changed the political balance throughout Western Asia, which in those days was the civilized world. In that space of time the Mycenaean Greece that Homer described gave way to the cultural glory of Classical Greece, Assyria and Babylon rose to be regional or world powers and then collapsed; Assyria in 612 BC under invasion by the Medes from the east aided by the Babylonians, and Babylonia itself only seventy-three years later, in 539 BC to the Persians. The Hittite Empire in Anatolia had already succumbed to foreign invasion about 1190 BC after a period of 490 years as a great power.

Egypt had awakened about 1565 BC with the New Kingdom from its long period of isolation. That ended with Ramesses XI in 1085. Thereafter it sank into relative obscurity under a series of ineffectual pharaohs till it was conquered by Alexander the Great in 332 BC. After that Egypt was ruled by a succession of Greeks till in 30 BC it became a province of the Roman Empire. The next native Egyptian to rule Egypt was Gamal Abdel Nasser in the 1950s of the second millennium CE. That does much to explain his popularity in Egypt.

The period started with the emergence of horse-drawn fighting vehicles from the grassy vastness of the herders of Siberia into the civilized world of the Middle East. The horse did not make an immediate appearance in the Middle East because it was not a native of that area. So people like the Sumerians had to yoke their battle wagons to the wilder and less tractable onager. But by half way through the millennium the domesticated and tamed horse had arrived in Babylon and Assyria, and also in Egypt and the Palestinian countries that were the scene of Egypt's new forward foreign policy. From now on warfare became a test of the control of masses of fast moving vehicles. The influence of the newly arisen countries, Egypt, Hatti, Mitanni and Assyria depended on the number of chariots their army commanders could produce in the field. These chariots could be numbered in the thousands. They must have been exceptionally impressive to behold as they galloped and bounced through the dust storms of their charging front ranks' passage.

For the horse was still a draught animal. When it was ridden, that was by a few scouts or messengers sitting on a blanket. With the invention of the saddle still in the future, riders' security on the animal was still not firm enough for them to be able to fight as cavalry, in drilled coordination as members of a body of other riders. The Assyrians managed it in the later years of their empire about 150 years before its end with horse archers sitting on blankets or fleeces without a saddle with a tree. But they were an exception and no other contemporary army seems to have taken up the idea.

The 2nd millennium BC was the period of the rise of growing international powers who fought what they saw as defensive wars against each other to protect their present or possible future interests, which at that time were usually the securing and protection of trade routes. If these wars often became aggressive, that was in the nature of politics and of war where attack is often the surest defence. The motivation, nature and degrees of success of these wars will be examined in greater detail under the sections on the individual countries. At the same time as the major powers were fighting each other with armies whose members were often not natives of the metropolitan country for whom they were fighting, they were being attacked by hungry wilder men, barbarians if you like, from less developed areas on the major powers' borders, whom we think were possibly horse riders. These invaders were determined to get access to the more fertile land that the settled societies occupied. So as well as aggressive wars against other organized states to maintain control of trade routes, settled societies very often had to fight defensive wars to keep the barbarians out. But in the case of the Assyrians' enemies the wild men very often lived in mountainous countries, which was much easier for horsemen to cross than it was for wheeled vehicles. And this is what I have begun to think was what gave the Assyrians the impetus to develop cavalry ahead of other countries that were in the plains that were more 'Chariot country'.

In his powerful book, *On Killing*, Grossman, who is a United States infantry officer and also a psychologist, makes important comments on the significance of the chariot in warfare.

He deals with the aversion of all soldiers to killing a man whom they can actually see and the efforts of all armies to overcome this. He sees the chariot as the first team-managed weapon. This would divide the personal responsibility of the members of the crew for what they were doing among its individuals and therefore reduce the moral responsibility of each member of it for the killing that was the result of their joint efforts.

Why highly militaristic empires that seem to be financially secure should suddenly collapse presents great problems for historians. The fall of the Roman Empire is a well-known case in point. There are numerous reasons, and as many interpretations. The problem is to rank them in importance. All too often all we have is the report of some major defeat. But that is never enough alone to satisfy the historian. And the collapse of the great Middle Eastern empires is not as well recorded as that of the Roman Empire.

The most obvious reason could be military defeat. But there is also the possibility of financial collapse. In the 18th century CE, in the days of Gibbon's *Decline and Fall of the Roman Empire*, moral collapse was a popular explanation.

We are not so attracted to that nowadays. But in their declining days large empires do have problems attracting to their armies young men who are willing to risk death in the hope of future benefits, land, loot or bonuses, arising out of being on the winning side. By the end of the Assyrian Empire we are fairly sure that many Assyrian soldiers were conscripted from former conquered peoples, and were not native Assyrians. That was certainly the case with the later Roman army and it has been the case with the armies of over-stretched economies ever since.

There was certainly a change in military equipment and its use in Western Asia around the beginning of the second millennium BC. The chariot came into use as the new prestigious and powerful fighting vehicle in the area.

We have little information for the beginning of the second millennium BC on military fighting vehicles in Western Asia in the period after the heavy Sumerian onager-drawn battle wagons of some 900 years before. Then the chariot became the fast fighting vehicle that every state that aspired to greatness had to possess. It seems as if the spoked wheel, a Siberian steppe land invention, was the innovation that replaced the heavy solid wheel and with the importation of the horse, a native of Siberia, into Syria and Mesopotamia came the innovations that made warfare mobile.

The number of chariots that an ambitious country considered essential to ensure victory and maintain prestige went from little more than a hundred in the 17th century BC to several thousand over the next six centuries, even if each carried little missile power for the expense involved.

To our regret we do not know the tactical use of chariots in the Middle Eastern Bronze Age. But it was probably much the same as that of later cavalry, of which we do know quite a lot. The chariots were most likely used for skirmishing and flank attacks after the enemy had been 'softened up' with arrows. But what must be included in any assessment of the wealth of any Bronze Age power was how fantastically expensive a chariot arm was.

For horses are susceptible to leg tendon injuries and a two-horse vehicle would require a third spare replacement horse for any incapacitated in a collision with another vehicle.

The expense of having a chariot arm can be seen from an account in the Old Testament. King Solomon (970–931 BC) is said in 1 Kings, chap 10, verse 29 to have paid 150 shekels of silver at a time when a shekel weighed 11 grams or 36 troy ounces for each of his chariot horses. His predecessor David (1010–970 BC) paid 600 shekels for each chariot. For comparison note that according to 2 Samuel 24.24 David bought a team of oxen and a threshing floor for 50 shekels. Earlier, in the book of Exodus in the 13th century the compensation for the death of a slave was fixed at 30 shekels of silver. In addition, Stuart Piggott has estimated that eight to ten acres of grazing land would have been required to maintain a chariot team and the 'ground crew' of grooms, a small army of scribes, carpenters and veterinary attendants would have to be housed and fed. The compound bow that the chariot warrior used took five or ten years to make.

Also, keeping track of the chariots and charioteers necessitated a small army of scribes and clerks.

The carrying of the financial burden of the ownership of chariots varied from state to state. The aristocratic *maryannu* of Mitanni seemed to have owned their own chariots and teams and in Assyria during the long 'off season' between campaigns the chariot drivers took their horses home with them, although the state provided their fodder. In Egypt the charioteer provided the horses while the government issued him with the chariot.

Selecting a chariot team is a difficult and expensive business, which must start with more horses than will finally be required. For not only must the animals be of the same size and conformation but they must be compatible with each other. Horses are hierarchical animals and they have their share of bullies. You must be prepared to discard the unsuitable ones you have paid for earlier.

The numbers of chariots that appear in the annals and other accounts of victories could well have come from optimistic estimates. Every experienced soldier knows the incidence of the lack of kit that makes the number of units on the inventory vastly greater than the number that are in condition to take the field. Minoan tablets from Knossos show that of twenty charioteers, that is drivers, only six had all the equipment necessary for action. One had horses but no vehicle. Another had a vehicle but only one horse, while another had a vehicle and two horses but no defensive armour. So only 21 per cent of the number on the inventory, the larger number on which the later write-up of the operation would be based, were fit for action. The problem cannot have been peculiar to the Minoan army, for the problem has plagued every army ever since.

There are manoeuvres every cavalryman knows for cutting his way out of the press of a scrimmage when he is brought to a halt. A charioteer cannot use them. He cannot spin his vehicle on the spot. He has to keep moving forward.

A possibility has been considered that chariots were principally used as battlefield taxis for infantrymen who then dismounted and fought on foot rather than using the chariots as fighting vehicles. But it is the experience of those who study the shifts made by the quartermasters who have to provide the tools with which wars are fought that if a piece of equipment can be used in action it will be.

All wars since antiquity have gone through a period of ineffectiveness, inefficiency and incompetence in the early stages as pre-existing equipment and methods have proved inadequate. There is no reason why that should not have applied in the second millennium BC. As Field Marshal Rommel said, the side that makes the fewest mistakes wins.

As wheeled vehicles, chariots were always limited to operate in an area where there was level ground free of obstructions or even small rocks for to hit them could overturn the vehicle. In fact, driving an unsprung vehicle at speed even over absolutely level ground is such a bumpy business that if you lean on the vehicle body for support you see double, one above the other, as your eyeballs vibrate.

The evidence is that even the light fast Egyptian chariots were found to be so vulnerable to attack by infantry that each had to be accompanied by 'runners' who accompanied it on foot to defend it.

Any consideration of the efficiency of military fighting vehicles must also take into account the quality as soldiers of the men who manned and serviced them. There was no general conscription in the ancient Middle East, although we know that in Egypt there was conscription of a proportion of temple servants and in the Hurrian town of Nuzi there were men with civilian trades who were called up as soldiers, but it does not seem to have been general conscription. The infantrymen seem in general to have been professionals, supplemented where necessary by tribal or allied levies. Campaigns were undertaken in a short season between harvest and planting. After that the majority of the men returned home to their regular means of sustenance, farming.

Of course we do not know the standard of training that was demanded of them. It is extremely doubtful if they would have been a sergeant major's delight marching in step and swinging their arms smartly with split-second precision as if on parade. But their weapon handling would have been of a high standard. For they were mostly of a farming background, used to handling tools and indeed weapons in days when every man was his own personal defence. They would not require training in that.

A more important question for the historian is the ability of ancient empires to field armies recruited from recently conquered peoples. The presence of soldiers in an army who do not come from the people of the metropolitan power has often been claimed as being an indication that the quality of the army had declined. But army commanders have always been aware of the problem of making soldiers fight who have no emotional attachment to the cause for which they have to fight. Generals have depended, with considerable success, on tough non-commissioned officers to instil cohesion and a willingness to go forward in troops who would rather give up. Soldiers in general fight from loyalty to the few comrades on either side of them whom they know rather than from the lofty sentiments of patriotism beloved of politicians.

It appears that from the Early Iron Age the chariot lost its importance in the battlefield, being replaced, at least in the Assyrian army, by cavalry. In other armies it seems to have been replaced by more infantry. But we have an overwhelming amount of evidence of what the Assyrians were up to from the reliefs that lined the walls of their palaces. We do not have this for other countries.

It should be pointed out here that as we do not know the ancient names of many pieces of saddlery and equipment connected with chariots and horses; the names of modern items of similar appearance and function are used. These are often very different from their ancient counterparts.

Examples are the snaffle bits and the bridle nose bands. Ancient snaffle bits had circular or rectangular side plates or cheek pieces that do not feature on modern ones and have no modern English name. These I call cheek pieces, a term that is now used for the short straps by which the bit is suspended from the horse's head. Crouwel calls these cheek straps, but cheek pieces is the modern English term for them among riders, so I have called them cheek pieces as well, while trying to make clear to which kind of cheek piece I am referring.

The band that goes up and over the horse's nose on an Assyrian bridle I have called a nose band because it is in a position close to that of the nose band on modern bridles. But the Assyrian one is not actually on the nose, being higher up the face, and it must have had a different function, if it had any other than ornamentation,

The chariot was showy, an indication of wealth and power. It was probably less effective on the battlefield than its expense would have merited. But it was an example of the peak of technology of its day and inspired respect and awe, like the nineteenth-century battleship. A country that had them was a major power. I contend that therein lay the reason why this impressive vehicle, which in fact was something of a white elephant, was kept in its thousands when the military advantage of it did not from a technical point of view justify its expense.

Riding a horse without a saddle is not all that difficult. But fighting from the back of an excited horse in close company with other horseman, leaning over and hitting someone without a saddle to anchor you is a very different thing. The Assyrians with their horse archers went a long way towards that and were the first in the field.

Truly the dawn of the horse warriors had broken.

Dislikes

- Difficult to follow at some points, lack of cohesion from thought to thought
- Lack of punctuation
- pg XXVI, does not specify the # of horses that would make up a chariot team → difficult to grasp the grazing land (just 1 chariot or all the chariots)

Likes

- Provided clarification b/w "domesticated" & "tamed" → no confusion
- Addressed the limitations of primary sources → best descriptions come from

Likes (con't)

Assyrian & why Assyrian examples are the majority of sources

- Provided his definitions used in his book
- Provided a general timeframe for warfare (during Spring/summer → accommodates grazing for horses used)
- Differens b/w solid wheels (humanitarian, religious, ceremonial) & spoked wheels (war)

Chapter One

The Domestication of the Horse

It is arguable that the first and greatest leap forward in the struggle by mankind (*Homo sapiens*) to control his environment was the domestication of the horse (*Equus caballus*). A case could be made for the argument that man's use of the horse did more to extend human influence than did steam, oil or petroleum. The horse, whose native homeland is the arid steppe of Central Asia, provided mankind at one go with food in the form of milk and meat and shelter from the elements through its hair and skin. As a draught and riding animal it went on to extend the range of human environment from a few miles to tens or hundreds of miles.

The horse also allowed *Home sapiens*, that most aggressive of species, to fight more efficiently to retain or extend the area inside which it could dominate others of its own kind. When that aggression was organized on a large scale as chariotry and cavalry it also made resisting rival human groups' threats much easier to achieve.

There are several species of the genus *Equus* that have existed. Apart from the horse, of which there were by the most generally accepted count, originally three wild sub-species. After that there were the half asses, the onagers, the donkey, and another three sub-species of zebra. The horse proved the most suitable through domestication and taming for mankind's use.

Here I should expand on the difference that I have already touched on in the Introduction between the ways in which I use the terms domestication and taming and those in common use. In general, domestication is used to describe the training of animals to do basic things that people want, such as to stand still when required to and to allow themselves to be handled and made to work at man's convenience. But because in a subject like this we shall be looking at people who may milk horses and use them for meat but apart from that do not to our knowledge ride them or use them as draught animals, I shall take domestication to be the accustoming of a zoological species only to living with humans, being handled by them, and being moved around by humans. Any further bending of the animals' wills to letting someone harness them or getting them to herd other animals at command I call taming. The onager was at times domesticated, but it never proved sufficiently tameable to make it a safe and reliable working animal. The horse may be both domesticated and tamed.

It will clear up a lot of confusion if we start by looking at how the horse and its relatives the half ass, the ass, and the zebra are all connected to each other. They are all members of the zoological order Perissodactyla, odd-toed hoofed animals, along with, unexpectedly, the rhinoceros and the tapir of south-east Asia and Latin America, while cloven hoofed animals like the cow belong to the order Artiodactyla.

The horse and its equid relatives are all members of the genus *Equus*. Members of a genus do not breed with each other in the wild and if they are induced to do so in artificial conditions like in a zoo or to produce crosses like a mule for human convenience they produce offspring that is almost always not fertile.

Below the genus *Equus* are several subgenuses, or subgeni if you are a Latin purist, which each contain several species. The first is the subgenus *Equus*, which contains all the known species, or races, of horse such as Przewalski's horse and the other early species of horse that have now been merged through human cross-breeding programmes. Members of the different species of the same genus can breed with each other and produce fertile offspring.

A second subgenus of the genus *Equus* is *asinus*, the asses. It contains three species, the kiang or Tibetan wild ass, the onagers or half asses and the donkeys. One sub-species of onagers is the Mesopotamian onager *Equus hemionus hemippus* that the Sumerians used as a draught animal, which is now extinct. Another species is *Equus asinus africanus*, the African wild ass, from which the donkey is descended.

After these comes the subgenus *hippotigris*, the common zebras. There are three species of that of which one, the quagga, is now extinct.

Finally there is the genus *dolichohippus,* which contains one species, Grévy's zebra.

The natural habitat of the horse is the long strip of grassland that stretches across Asia from the *puszta* of Hungary in the west to Mongolia in the north of China in the east. After Hungary this grassland or steppe stops, with a break for the Carpathian Mountains. It resumes in Moldovia to the east of the Carpathian Mountains and then crosses the Ukraine. It skirts the southern tip of the Ural Mountains and passes through Kazakstan, Tajikistan and Uzbekistan before reaching Mongolia. In summer the temperature reaches 104°F (40°C), while in winter it can drop to -40°F (-40°C). In the West it is only a few hundred miles in width from north to south but it becomes much wider to the east. To the south it passes past the north of the Caspian Sea and the Tibetan plateau, while in the north it reaches the taiga, the pine, spruce and larch forest that fills the northern part of Siberia as far as the Arctic Ocean.

The horse cannot be wintered out north of 52° while equally it is not by nature suited to the warmer climate of Mesopotamia. The onager was a native of Mesopotamia, Iran and parts of Pakistan, while the donkey was originally a native of the Sudan, although it is now generally spread throughout the Mediterranean lands and the Middle East.

Views differ on how many different kinds of native horses of Eurasia occupied Europe and Asia after the last Ice Age when the American horse had either died out or crossed over the dry Bering Strait into Asia. It is from them that the modern breeds of horses are derived and varied schemes have been proposed to explain how this happened. The minimum number of proposed possible wild ancestors of the modern horse is three and if that plan is not as exact as some of the others, it is easier to comprehend. They are the extinct European pony (*Equus abeli*) of western Europe and Germany, the extinct tarpan (*Equus ferus silvestris*) of eastern Europe, of Poland and the Ukraine, and the still living Przewalski's horse (*Equus ferus przewalskii*) of Central Asia. The European pony is long extinct as a separate sub-species. The last tarpans were exterminated in the Ukraine in 1850 because of the damage they did to crops and the stallions' predilection for domesticated mares. It is not known if any of Prezewalski's horses are still roaming the steppe of Mongolia in a natural wild state. The last ones that were seen for certain in the wild were in 1966, but at least a hundred are alive and breeding in various zoos. They are named after the nineteenth-century Russian, in spite of his Polish surname, geographer and explorer Colonel Nikolai Przewalski (1839–1888), who was the first to describe them.

The northern border of the steppe is the limit north of which the horse cannot survive in the wild naturally without being artificially fed, pastured or stabled. This is the southern border of the taiga and its terrain of forest and swamp also happens to be the southern natural limit of another animal that was an early candidate for domestication in Post-Glacial times. This is the reindeer, known in North America as the caribou. That may even have been domesticated and tamed before the horse. We do not know, but it is possible. It is highly specialized for survival in the taiga and adapted to the terrain. The reindeer lives on a specialized diet of a lichen called reindeer moss (*Cladonia rangiferina*) that grows in the taiga and its splayed hooves are soft in summer and do not sink into the swamps as the hard hooves of horses would. In winter the edges of its hooves become sharp and hard and cut into the ice so that it does not slip, something that horses are terribly prone to on ice. Other animals left over on the taiga from the Ice Age such as the black bear and the Siberian tiger do not prove amenable to domestication, although a few other smaller species such as hawks and ferrets do.

Reindeer are best known in Britain and the United States as the mythological animals that draw Santa Claus' sleigh at Christmas but they are still domesticated animals that are herded by the Lapps of northern Norway and Sweden and the various Turkic speaking tribes of northern Siberia. They are milked and some of those peoples use the reindeer as draught or pack animals, while lightweight members of the communities ride them.

In modern times the Turkic-speaking tribe of the Uryanchai of the Sayan Mountains of Mongolia ride reindeers and put pack saddles on them. They have rather primitive riding saddles, which do not have a conventional tree as modern riders in Europe and America understand the term. Instead horizontal wooden spars rest on pads on either side of the animals' spine and between them at back and front are X-shaped forks from which is suspended a leather seat. Unlike the Lapps they do not use sledges and they have no dogs.

If reindeer were the ideal animal for domestication in the taiga, the same was true of horses in the hard grass of the steppe. For they do not thrive on lush grass, although further south they eat it with relish and are inclined to contract laminitis, inflammation of the hoof, from the excessive protein in it. Other nomadic pastoralists of northern Siberia such as the Tugus and the Yakuts also ride reindeer and the supposition has been expressed that riding on horses in the more southerly grassland of Central Asia may have been adopted in imitation of the reindeer riding of the north. But to be honest nobody knows for certain. In this field of enquiry so much must remain supposition.

The earliest evidence for the domestication, as distinct from the taming, of the horse comes from various sites in the steppe in what can be regarded as either eastern Europe or western Asia. The problem in attempting to estimate the state of domestication of horses is that it does not produce the skeletal changes in them that it does with animals that are kept and bred for meat, where larger and larger muscles demand heavier bones. Horses kept in captivity do basically what comes naturally to them. I have seen a thoroughbred ex-race horse do a very good canter pirouette, an advanced dressage exercise, out in her field out of sheer high spirits. She certainly could not do it to command with a rider on board.

But there are several methods by which the estimation of the date of the domestication of horses has been attempted from examination of the skeletal remains. These often have to be based on the examination of fragments of bone separated from any close-by dateable archaeological context. So often the archaeologist is working with partial information.

One is the Size-Variability method. This works on the assumption that animals that are kept in pounds, paddocks or stables, would vary more in the size to which they grow as calculated from the width of the shaft of their leg bones than would horses that lived unrestricted in the wild. This works quite well with the examination of cattle and sheep, which show much greater increase in size after domestication than do horses. But the shortcoming of this method is that we do not know the exact methods of management employed by people who were involved in the early domestication of horses. It does not work with the horses domesticated by the Native Americans because they did not keep their newly captured feral horses in paddocks but hobbled them so that they could walk but not run and were free to find their own fodder. Thus these animals do not show the variation in size compared with wild horses that would be expected. The Size-Variability method produces an estimated date for the domestication of the horse of 2500 BC.

A second method for the estimation of the degree of domestication of horse remains is the Age at Death method. This is based on the assumption that animals selected for slaughter from a domesticated herd should be of different ages and sexes from those of animals obtained from hunting. A domesticated herd would contain fewer males than a wild herd and this is easily identified from dental examination. Females will be kept in greater numbers for breeding. But the sex component falls down when it is appreciated that in a herd of wild horses one stallion safeguards a number of females and will fight other stallions to keep them away from his mares. In this way his genes will be perpetuated, although his attitude to sex is not so eugenically inclined.

A hunter, like a predatory animal, will tend for ease to take the young of any prey animal as they lag behind the herd and become separated from it. A herdsman or butcher will select older animals for slaughter as they have more meat or are past the age of breeding.

There is of course the date of death estimation by Carbon 14 (C_{14}). But that only gives us the approximate date when the animal died, with a large plus or minus that is not terribly helpful. It does not tell us what kind of life the animal led before it died.

Marsha Levine of the Cambridge McDonald Institute of Archaeological Research has studied age and sex data from two early Eneolithic (Copper Age sites between the Neolithic and the Bronze Age) that contain the fragmentary skeletal material of horses in the steppe. They are Dereivka in the Ukraine and Botai in northern Kazakhstan.

Dereivka is 400 kilometres south-east of Kiev, on the border between the steppe and the forest to the north. The steppe is only 400 kilometres wide from north to south at the longitude of Dereivka but east of there it widens to a thousand kilometres wide north of the Aral Sea at Botai. The dating of Dereivka has been the subject of much controversy and revision. The C_{14} date obtained from it by the original excavator in the 1960s was 4200–3700 BC, making it the oldest site at which the remains of domesticated horses had been found. But C_{14} examination of another part of the site that contains what is interpreted as a cult burial of the head and hoof bones of a seven to eight-year-old stallion and two dogs gives a much later date, from the time of the Scythians in the 6th and 7th centuries BC. But whatever its date,

Dereivka certainly contains evidence of an early stage in the domestication of the horse.

The human settlement at Dereivka that is near the slaughter yard is a village that has beside it an area that contains 3703 pieces of the bones of slaughtered mammals, 2225 pieces of which, about 60 per cent of the total, are of horses. Marsha Levine has examined the remains from Dereivka and also from Botai and has concluded that the horses at both sites were wild. She found that at Dereivka the majority of the teeth found were of animals between five and seven years old and that fourteen of the sixteen mandibles were of mature males. That was not the butchering pattern to be expected from a managed population.

She suggested that the Dereivka hunters stalked and killed the stallions when they went to investigate the strange intruders into their herd. This, of course, is to assume that the stallions became inquisitive and did not immediately get the whole herd to flee. But we must credit the Dereivka hunters with being skilled at what they did, for their lives depended on it. And we do not know what ruses they adopted to encourage the stallions to venture within bow shot range.

Botai, on the Imam-Burluk River, a tributary of the Ishim River in Aqmola Province of Kazakhstan, is a type site, one that gives a particular culture also found elsewhere its scientific name. The same culture is also found at the sites of Krasnyi Yar and Vasilikowka, which are dated to between 3700 and 3000 BC. At Botai, Levin found on the contrary that the age and sex profile suggested that whole herds of horses were slaughtered.

David Anthony and his wife Dorcas Brown have done much research into and study of the wear on horses' teeth by the bits worn by animals that are used for either riding or driving. This is the only wear that is detectable on a horse's skeleton by its being made to work in one of those capacities. They found that not only metal bits but also bits made from soft material such as leather or rope cause wear on the lower second pre-molars immediately behind the bars of the horse's mouth, that part that has no teeth in it, and on which the bit lies.

Botai is a site that has been reckoned to have been inhabited in the Chalcolithic Age at the beginning of the Neolithic by a group of specialized hunters who rode horses as well as hunted them on foot. This is a peculiar kind of economy that is known to have existed between 3700 and 3000 BC and only in the steppes of northern Kazakhstan. Elsewhere such sites have been known to have contained fragments of horse bones that amounted to 99.9 per cent of the osteological finds. Botai bears this out. Of the 300,000 fragments of bone found there, 99.9 per cent of them were those of horses. Anthony and Brown visited Botai and examined the second pre-molars of excavated horse teeth to look for the typical wear on the enamel caused by bits that are worn by horses that are not only domesticated but tamed for driving or riding. Of nineteen teeth examined, five showed wear consistent with their possessors having been mouthed for a bit. The others did not show sufficient wear to justify a conclusion that they had worn a bit. There is disagreement among osteologists as to whether the Botai horses were domesticated and kept in some kind of confinement or hunted in the wild as required for work. All the arguments depend on assessments of how the Botai people lived. The teeth do not give the answer to that and the Botai people, as illiterate nomads, have left no records. But it does seem probable that some of the horses were ridden while others were not. It is unlikely that the horses with tooth wear were used to draw wheeled vehicles. These would not have been useful for hunting wild horses. But if these suggestions are correct, they show that by the 3rd millennium BC people had started to ride horses.

The self bow, made only of a single piece of wood, was not powerful enough for hunting a fast moving prey until it was made too large to be handled on a moving horse. That had to wait for the invention of the compound bow whose elasticity was increased by gluing strips of horn and sinew to it. At the same time it became small enough to be shot by a skilled archer who was also an experienced rider. That did not come about till about 2000 BC.

Much has been made of the suggestion that the development of horse riding also brought about an increase of warfare. But that demands a sociological change where people agree to subordinate themselves to a leader and follow his plan. War demands the massing of superior force, of men acting together to overwhelm the opposing force by their combined action. Individual daring acts by a fearless rider lead rather to posthumous fame. Organized warfare had to wait for civilizition, which was far in the future for the nomads of the steppe.

The archaeological evidence indicates that during the Eneolithic, about 4000 BC, there was raiding from the steppe westwards into the settled agricultural areas of eastern Europe, into what is now Transylvania. We think this was by people who rode horses. This was not mass invasion like the incursion by the later Huns, but the raiders can be deduced from their burial customs to be newcomers to the area.

Societies are to this day very conservative about their burial customs and archaeologists therefore study these carefully. When a new way of burying the dead appears in an area, we can conclude with reasonable confidence that there has been an ingress by a new foreign group of people. That was the case when people belonging to the Sredni Stog culture appeared in western Asia and eastern Europe in the 4th or late 3rd millennia BC. The Sredni Stog people buried their dead in individual graves marked on the surface with a circle of stones, quite different from the burials in grave pits of the earlier inhabitants of Dereivka, which were not marked on the surface. The Sredni Stog bodies were placed in a most distinctive position, lying on their backs but with their knees raised and their legs bent up sideways. Their skeletons were also distinctive, being much finer than those of the earlier Dereivka people, and the skulls were narrower than the heavy broad faces of their predecessors.

Sredni Stog, which is an island in the River Dnieper north of the Sea of Azov, gives its name to this culture, a group of people with distinctive customs and possessions, which is dated to between 4200 and 3800 BC. Its members were present at Dereivka, and their remains were excavated at a higher level and therefore later than the people who butchered so many horses and in whom we have been interested so far.

It will cause the reader no surprise to learn that there is considerable controversy among archaeologists as to the significance of this site. For with this very early period we are not dealing with historical archaeology as we are with the Assyrians, Egyptians and Hittites of the Middle East, where the pottery and skeletal evidence is well supported by pictorial reliefs illustrating the technology of the period and where there are contemporary written accounts, which, however much they are 'spun' for political reasons, give us a version of what was happening,

Sredni Stog settlements contain twice as many horse bones as are found at earlier settlements in the Dnieper valley. We cannot be sure why this is.

It could be because the climate was getting colder in 4200 to 3800 BC and it was easier to keep horses than cattle and sheep in snowy conditions. But the percentage of bones in Sredni Stog sites that come from species other than horses varies from place to place and the reasons for this could be economic rather than caused by climate.

We cannot be sure that the Sredni Stog people who buried their dead at Dereivka actually rode their horses. But it looks as if they did. For they were highly mobile, taking part in trade over long distances and the stone weapons found in their graves were of higher quality than those of the less mobile settled people with whom they associated, and whom they raided round about 4200 BC, when their remains have been found in the Danube delta west of the Black Sea.

About this time, around 3800 to 3300 BC, there were other significant population and cultural changes in the European-Eurasian border to the west of the Black Sea, which are to be seen in shifts in material prosperity in the Caucasus.

Here Sintashta is an archaeological site that has had a profound impact on our understanding of the move over from mounted nomadism to chariots in the steppe between 2900 and 1780 BC. It is a fortified circular town of a type that we now know is to be found elsewhere in the northern steppe east of the Ural Mountains. It lies immediately east of the Urals and north of the Caspian Sea on the border between the steppe and the forest.

This general survey of the story of the invention and development of chariotry and cavalry must now be brought to a close as far as the domestication of the horse is concerned. For the story now carries on with the horse well domesticated and tamed while further south in Mesopotamia, where it is not a native it is the onager that was used as the available draught animal until the middle of the 2nd millennium BC.

Archaeology has been described as 'An art that uses scientific methods' and while we try to present it as a logical progression to greater and more certain knowledge, it in fact goes ahead by fits and starts, leaps and bounds, depending on 'The accidents of excavation'. Excavation of a site is only decided on after lengthy and careful consideration of the probable information on an archaeological problem that could be obtained there. The actual information uncovered may be very different. For it is only after you start digging that you discover what is really there. It may be of great scientific interest but it may be rather different from what had been expected.

Many of the conclusions that an archaeologist has to reach regarding the age or significance of a find, which is often only a subtle change in the colour or consistency of the soil, have to be based on partial evidence that is not as definite as he would wish. He is dependent for his conclusions on the experience of similar situations that he and his colleagues on site have observed in previous occasions. If archaeology is a science it is not always an exact one.

Chapter Two

Central Asia

The Central Asian steppe, as already outlined, is an immense harsh grassland 5000 miles (8000 kilometres) long from east to west and 600 miles (1000 kilometres) from north to south. It stretches from Manchuria in the east to Hungary in the west and is bordered in the north by the *taiga*, an extensive band of forest, and on the south by the Gobi and Takla Makan deserts and the Tien Shan and Pamir mountain ranges. In modern political terms it occupies the Chinese provinces of Inner Mongolia and Sinkiang and goes on westwards to cover Russian southern Siberia and Uzbekistan, Turkmenistan, Kazakhstan, Kirghizstan, Tajikistan, parts of Afghanistan and Iran, and ends up in Hungary.

The Central Asiatic steppe is a land fit for nomads; and only nomads. Years ago scholars, and those who wanted to improve other peoples' worlds, considered that nomadism was more primitive as a way of life than settled agriculture. Opinion has changed now. In a land with only one low yield resource that is widely dispersed, as is grass in Central Asia, nomadism is an intelligent way of life. When the resource is depleted, the grass is grazed down, or you are threatened by a new group of interlopers, you move on.

This was the way of life in Central Asia in the 2nd and 3rd millennia BC It still is, in spite of encroaching modern industry. And the nomads had a most valuable means of moving themselves and their possessions to pastures new – the horse. The domestication and taming of the horse started in the animal's natural homeland, the Central Asian steppe, and from there that most useful animal spread over the rest of the temperate zone of the world.

So it is hardly surprising that the domestication of the horse originated in Central Asia, even if wheeled fighting vehicles did not appear for thousands of years later and then in another area, Sumer in southern Mesopotamia, and those were battle wagons, not chariots. And they were not drawn by the horse but by a different species of equid, the onager. In the Carpathian Basin,

Transylvania and Slovakia, numerous pottery models attest to the use of disc wheels, and therefore vehicles, around the middle of the 2nd millennium BC. Among the pottery wheels is a group that represents spoked wheels, with four spokes. C_{14} dates of spoked wheels in Slovakia are 1400+ to 200 BC. It is not known whether they come from two- or four-wheeled vehicles.

Actual remains of two vehicles dated from 1300 to 1100 BC with spoked wheels were found in the water-logged tombs in Barrow numbers 9 and 12 at Lchashen on the shore of Lake Sevan in Armenia, which is mountainous country south of the central Asiatic steppe. They were two-wheeled vehicles and were considered by many to be chariots, although Piggott disagreed because it would have been dangerous for the crew to stand to fight from them while they were in motion. But we do not know if Chinese archers had to stand to shoot their arrows. They might have been accustomed to kneeling or sitting or squatting.

In addition, fragments of a single spoked wheel of a different vehicle that was not itself found, were discovered in Barrow 10. A description with measurements was given of Chariot 1 from Barrow 9 by the Armenian excavators. It is virtually identical to Chariot 2 from Barrow 11 and can be taken as typical of the Lchashen chariots. It had a light body 1.10 m wide and 0.51 m deep in from the end. It is slightly curved behind with a rail on slender uprights at the back and sides. Slots in the frame suggested a floor of interwoven leather straps. The axle was central and 2.25 m long. The wheels rotated freely on it. They were 0.98 m in diameter on Chariot 1 and 1.02 m on Chariot 2. They had turned naves 43 cm long into which were mortised 28 spokes, which were fixed at their outer ends into a felloe of two half circles of bent wood with scarfed joints. The felloe was probably originally 4 to 5 cm thick with a tyre of wood or leather. The draught poles were missing and the straight pole in the drawing is a reconstruction subject to some criticism.

In Barrow I the vehicle had a body with a rail across it at the front and it was open at the back. Its wheels had eight spokes while the chariots in Barrows 9 and 12 had six spokes on their wheels.

The Lchashen ones were not the only early spoked wheels found in the Caucasus. The remains of two light wheels of 90 cm to 1 metre in diameter with 10 spokes were found partially buried upright to a depth of 30 cm in the floor of a grave in a cemetery in the southern Urals, on the Sintashta river near Chelyabinsk. They were dated to the early 15th century BC. The track of the undiscovered vehicle to which they would have belonged is estimated at 1.25 to 1.30 metres.

The dimensions of the two-wheeled vehicles from Barrows 11 and 9 were:

	Chariot 1 from Barrow 11	Chariot 2 from Barrow 9
Wheel diameter	0.98 m	1.02 m

Below are given dimensions that were the same for each chariot. The hubs were 43 cm long, remarkably long for the hubs of any wheel, and revolved on an axle 2 m long. The wheels had wooden lynch pins.

The spokes were 2 cm wide and 1 cm thick and each wheel had 28 of them.

The felloes were made of two half circles of bent wood 4–5 cm thick with scarf joints. There were no metal tyres but there was an outer strip of thin wood or leather on each wheel. The track of the wheels was around 1.10 metres and the axles were under the centres of the bodies.

The body of each chariot was 1.10 m wide and 0.51 of a metre deep, straight in front and slightly curved outwards behind. It was fastened to the square sectioned centre portion of the axle underneath it by dowel pegs passing through small blocks or pads at the outer ends and at the centre it was attached to the draught pole, which was estimated to have had a length of at least 3.5 metres.

Oblong slots were cut in the framework round the floor of the body, Fifteen of them were cut along the front and back of the framework of the chariot floor and eight were cut at each end of the floor. These suggested to the excavator that the chariot had a flooring of interwoven leather straps and that round the floor was a railing that was carried on slender uprights half a metre high. These, he suggested, surrounded the body at the sides and the rear, leaving the front open.

There were no surviving individual yokes for each of the horses.

Three remarkable features of these Asiatic chariots that make them very different from Middle Eastern chariots are:

1. The open bodies are broader than they are deep from front to rear, with low railings at the sides and rear.
2. The number of spokes is greater than those found in the Middle East.
3. The felloes are made of two pieces of bent wood rather than the greater number of pieces carved in parts of a circle as in Middle Easter Chariots.

Bronze models of similar chariots were found in Barrows 1, 9, and 10. Horse bones have been recorded from more than one of the graves, although this is not a primary area of *Equus* domestication, which is further west in the steppe and forest-steppe zone of the Ukraine in sites such as Sredni Stog and Dereivka. Six rigid cheek pieces probably made from reindeer antler were found at Lchashen, dated by C_{14} to c. 4730 BC.

At Lchashen bronze bits are recorded. Six of them come from Barrow 2 and two from Barrow 8. They are variants of the type found at contemporary Caucasian sites in Armenia and Georgia. They have discs on their ends that are either openwork or wheel shaped, and that have spikes that press on the outsides of the horses' lips. Littauer (1969) calls them 'run-out bits' and they are used on driven horses that insist on running out to one side. They are not used on ridden horses. In riding, the problem of keeping a horse straight is dealt with by shifts in the rider's weight and the use of his legs. There are parallels with these devices in Bronze Age bits that are found in the Ukraine near Kiev, in the central Urals and in Romania.

The antler and bone side pieces might be considered as prototypes rather than derivatives of the bronze bits and are paralleled with other types found in Timber Grave cultures in Siberia.

The twenty-eight-spoked wheels of Lchashen vehicles come from a different tradition of wheel making from those of Egypt and the Near East. In the latter area the four-spoked wheels lasted up to the fifteenth century. After that there were as many as eight or nine spokes on some Syrian chariots. The Lchashen chariots were much less sophisticated than Egyptian ones in their use of bent wood and Littauer and Crouwel consider them as impracticable as fighting vehicles. But that would be to assume that they were served by crews who stood to fight as did those in Western Asia. If we think that they might have knelt as Chinese crews did, then the question of their serviceability as fighting vehicles would be open again.

Chinese chariots cannot be discussed without reference to finds at Lake Sevan. This idea occurred independently to Piggott, who dated the Chinese burials to between late Shang times around the 12th century BC to late in the Chou dynasty around 400 BC. Outside China the carriage from Barrow V at Pazyryk in the Altai of the early 4th century BC can be accepted as being of Chinese design.

Lchashen is close to the 40th parallel of latitude half way between the easternmost point of the Black Sea and the mid-point of the western shore of the Caspian. The route eastwards to China from there goes first south-east to skirt the southern end of the Caspian Sea and thereafter turns roughly eastwards. It is 6500 kilometres across rough country through high mountain passes from Armenia to where the Chinese chariots were found, from Lchashen to Anyang. That Chinese site is eight hundred kilometres south of Peking in south-eastern China and there was an established trade route from there to Western Asia. This was the Silk Road that in antiquity was followed by traders going to and from China and the Mediterranean. Along it was carried the trade in silk and spices to and from Central Asia and China. For most of its way it crossed country that was above 900 metres above sea level. It passed Tashkent, split to pass north and south of the cold Taklamakan Desert, which is of an oval shape 1000 kilometres long from east to west and 400 from north to south at its widest, and then eastwards into China. There is some evidence that the local climate was warmer in antiquity than it is nowadays, but it was still a very long walk. However, if you were the bearer of intelligence of a new kind of military hardware you could expect to be well rewarded when you sold it. For the general consensus of academic opinion is that the idea of these chariots went from west to east, from Central Asia to China, and not the other way round.

The Lchashen carts bear a close resemblance to the Chinese carts of the Shang dynasty, although they differ in details. They follow the same general tradition as regards wheels and bodies with row rails round them.

The principal difference between the Armenian and the Chinese chariots is that while the Chinese chariots have no enclosing rail at the rear, as one might expect on a chariot, the Lchashen ones have no protecting rail at the front. The Chinese carts are, in spite of their low guard rail, generally regarded as war chariots, although exactly how they were used in action is still a matter for discussion. And the Sheng dynasty and the Lchashen burials are of much the same date and where the Chinese got the idea and the design of their wheeled vehicles is still not known beyond doubt.

The presumption that the Armenian and the Chinese chariots are related has to be based on the close resemblance of them to each other. After all, how long does it take you to drive a light cart from Armenia to China or walk a few thousand miles with a highly saleable piece of military intelligence? Certainly not centuries.

Like those of China the Armenian chariots have large light wheels with numerous spokes. Like the Chinese vehicles, but unlike any other known war chariots, they also have bodies that are considerably wider than they are long and the rails round their sides are remarkably low, at half a metre possibly even lower than the Chinese ones. But a significant difference is that while the Chinese vehicles have no wall on the body at the rear, the lack of which is a further justification for calling them chariots, the Lchashen carts have no wall restraining the driver and passengers from falling out of the front of the body under the horses' hind legs. So whether this Armenian and Chinese pattern of vehicle is a fighting one that justifies our calling it a chariot or is only a processional one for use on prestigious occasions has been much debated. But we have Chinese written texts on war that mention the tactical use of chariots in action. They are dealt with in more detail in the chapter on China. There, they make it clear that the Chinese of the Bronze Age used two-wheeled chariots in which the driver, an archer and a halbardier sat, and the two warriors fought from a sitting position. So I feel we are entitled to call these Chinese versions war chariots and, I think, we can take the Armenian ones as being fighting vehicles also, even if they were not being used in warfare on the day of burial.

I feel that the interpretation of the Sumerian two-wheeled vehicle of which a bronze model was found at Tell Agrab is more complicated than that. We do not know its purpose. With no useful carrying capacity beyond the driver it would have been of no use for war or agriculture. And there is no recorded instance other than the grandiose and unrealistic wall paintings of Egyptian pharaohs driving their chariot while shooting arrows where a chariot did not have at least one archer aboard it as well as the driver. In the Sumerian astride cart the driver could not have driven it as well as fight from it. It was either a prestige vehicle, a prototype to show what a Sumerian city could do, or else the only weapon was the span of four onagers and local defence was provided by infantry accompanying it. If it was the city that stood in the site now known as Tell Agrab that held what is now known as an arms fair, there is no evidence that any other city acquired one of those vehicles. So my conclusion has to be that it was a dead end prototype that was not developed further.

The four-wheeled Sumerian battle wagon illustrated on the Standard of Ur is definitely a military vehicle. It will be dealt with in greater detail in the chapter on Sumer. The name 'battle wagon' has now come to be generally accepted for it. The two-wheeled Sumerian vehicle that the driver sits astride has as yet no generally accepted name. One that has been applied to it, and of which I approve, is 'astride cart' and is the one I use.

The terms chariotry and cavalry also demand definition. A handful of arrow-shooting nomads riding in vehicles or on their horses do not constitute a chariotry force or a cavalry unit as I would define those. Chariotry and cavalry as I use the words are organized and disciplined units of an army working in planned cooperation.

Chapter Three

Wheeled Transport before the Sumerians

After the domestication and taming of the horse as a means of conveying a single rider while he was travelling, hunting, and herding had been achieved in Central Asia, the next stage in the exploitation of the animal was to use it to move a wheeled vehicle. So now we have to search for the earliest example of the existence of the wheel. That is not to say that the society, call it a culture, that has left us the earliest known wheels were the ones who made that fundamental innovatory discovery. All that we can say is that they are the first we know of at present. And dates of finds from around 3000 BC are so inexact that we cannot possibly award the palm for being the first to invent the wheel to any one particular society.

So the next archaeological site that helps us in our study of the earliest chariots is Starokorsunkaya situated on the Kuban River in the Caucasus. It contains a burial mound made by a people belonging to the Maikop culture, who thrived round about the end of the 4th millennium BC. The people of the Maikop culture were great traders and we know that they had trading relations with the urbanized Uruk culture in Mesopotamia, which is dated to between 3750 and 3000 BC, before the full flowering of the Sumerian culture. And that is as chronologically exact as we can place them. In a Maikop culture burial mound, called by archaeologists Kurgan 2, soil traces of two solid wooden wheels have been found. There was no sign of the vehicle and we do not know whether it was two or four-wheeled, but the animal that drew it was most likely to have been the horse. This was around the time that we think that the Sumerians of the Uruk period went from using sledges to putting wheels under them and thus turning them into carts or wagons. But we cannot on the sparse and difficult to date evidence say who got the idea of wheels from whom. The horse was certainly not living in the wild in Mesopotamia in the Uruk period and the Sumerians had to use the much more fractious onager as a draught animal. Horse bones are rare in rubbish deposits in the Caucasus before about 3300 BC and afterwards become common.

We do not know exactly when the first wagon appeared on the Eurasian steppe. But a cup found at Bronocice in southern Poland that was dated to 3500–3300 BC has an impression of a wagon on it.

And large numbers of wagon burials of the Yamnaya culture have been excavated at sites on the Kuban steppe that have radiocarbon dates between about 3100 and 2200 BC. Their wheels, which all revolve on fixed axles, are made of two or three planks dowelled together and cut in a circular shape between 50 and 80 centimetres in diameter. The body of a typical wagon was about 1 metre wide and 1.5 to 1.6 metres long. One vehicle had a box seat for the driver and the passengers or goods in the vehicle were protected by a tilt of reed mats painted with red, white and black stripes and curved designs. The matting was possibly sewn onto a backing of felt.

It is thought that the Yamnaya people were peaceable nomads with large herds and we cannot say whether when they appeared in the steppe between 3500 and 3300 BC, they came from the West, from Europe, or from the south, from Mesopotamia, or from the Volga-Don steppes in the east, although current opinion favours the east.

To this day the Kurdish horse breeding nomads travel immense annual distances between Iraq and southern Turkey in the routine of their annual migration. For nomads do not wander at a venture. They have regular repeated routes that they follow in search of grazing for their stock.

Since the earliest times the menace of invasion by nomads from the wilderness has terrified settled agriculturalists. The biblical story of Cain and Abel, with its conflict between the pastoralist and the agriculturalist, is based on one such legend and that is not the earliest account of it known to us. There is an earlier Sumerian story written in cuneiform that recounts the same fear. The settled agriculturalists' received version of the reality of nomads was of masses of uncouth warlike horsemen pouring out of the east intent on the rape and pillage of the hard earned wealth of the settled area on which they were parasites. But that tale of savagery belongs rather to a later period of history, when the Huns descended on the faltering Roman Empire. The pastoral nomads of the 2nd and 3rd millennia BC practised an alternative to settled agriculture in an area, the steppe, where the climate and terrain would not support profitable agriculture and their nomadic neighbours and competitors were doing the same thing.

The remains of two wooden wheels were found in a grave in Plachidol in the north-east of Bulgaria. There were indications that there had originally been four wheels arranged at the corners of the grave around the skeleton. This is dated to c. 3000 BC.

Anthony's study of the origins of the Indo-European language sees a connection between Mycenaean Greece about 1650 BC and the cultures of the steppe and south-east Europe.

The finds in Mycenaean Greece include cheek pieces for chariot horses, specific types of socketed spear heads and masks for the dead, which are common in the Ingul River Catacomb culture of Russia and the Ukraine.

It must be taken into account that for archaeologists the term culture is used for a collection of people who use the same kind of everyday artefacts. These are often pottery buried in graves and indicate that there were many different cultures moving westwards into Europe in the Bronze Age. They moved slowly, pausing to take a cash crop each year to replenish their supplies. This was slow immigration, not speedy invasion at a gallop.

At least ten major culture groups inhabited the steppe and the steppe-forest region between the Ural Mountains and the Carpathian Mountains in the Middle Bronze Age between 2800 and 2200 BC.

Because they were illiterate, or to give them the benefit of the doubt, because they left no written records in an imperishable medium, we do not know any details of what the steppe peoples did with their wagons. Still we can claim that they were in advance of the Sumerians in the technology of transport. For they had lightweight spoked wheels when the Sumerians, who left much more for posterity in the form of written legends that were taken up by the Babylonians, had to make do with slow, heavy and inefficient solid wheels. And the steppe peoples' vehicles were pulled by horses wearing bits in their mouths, a technology that is still in use today, while the Sumerians had their battle wagons pulled by onagers restrained by a single rein attached to a ring through the gristle of the septum of the nose. I do not believe the alternative interpretation of the Sumerian inlay 'The Standard of Ur' that the ring went through the onagers' upper lips. For it would easily have torn out of them. But to excuse the Sumerians for their inferior draught animals, it must be pointed out that the horse was at that time unknown in Mesopotamia. We do not know whether onagers could have been trained to accept a bit. It has never been tried.

Thanks to the Sumerians' decision to write in cuneiform on imperishable clay tablets we know that they lived in comparative peace in their desert towns, although they did go to war with each other, mostly over water rights to the irrigated land outside the towns. Beyond that was E.DIN, the uncultivated desert between the towns, which gave its name to the later Semitic Eden in which according to the Book of Genesis humanity started.

Certainly the herding steppe peoples had towns like Sintashta, north north-east of the Caspian Sea, at the same time that the settled agriculturist peoples of the Uruk culture in Mesopotamia made their first efforts to record their philosophical ideas and little wars in writing. In the absence of written records we do not have equivalent records of life at Sintashta and other steppe towns like it. So we must depend on what we think nomadic steppe life would have been like and the pressures and influences under which they must have lived.

War leads to expenditure without limitation on new technology and it is a reasonable assumption that the constant movement to find new grazing in possibly unfriendly and unwelcoming lands made that portion of the residents of towns like Sintashta, who retained a herding life style, more attuned to war than were the urbanized Sumerians of the Uruk period. No doubt a good portion of the people of Sintashta were as urbanized as the Sumerians were, but the evidence suggests that the economy was dependent on herding rather than the irrigated agriculture that was the mainstay in southern Mesopotamia. So the steppe people advanced in the technology of transport to an extent that the Sumerians did not feel necessary.

It is concluded from the fortifications of palisades and ditches round their settlements that the people of the several Sintashta towns lived in fear of attack. The graves held more weapons than in earlier times. These included heavier arrow heads of a new type and larger spear heads.

Today most authorities credit the invention of the chariot to Near Eastern societies around 1900–1800 BC. Until recently scholars believed that the chariots of the steppes post-dated those of the Near East. The basis of this was the carvings or petroglyphs showing chariots on rocky outcrops in eastern

Kazakhstan and the Altai. These were thought to belong to the Late Bronze Age c. 1650 BC with the steppe cheek pieces later. But new evidence since 1992 has changed this view. This is the soil signs of chariots of the Sintashta culture and the Petrovka culture, which is dated as contemporary with the late Sintashta culture of about 1900–1750 BC in northern Kazakhstan. The wheels are buried in slots in the ground earth in graves rather than in the infill and only what was buried has been recovered. But they are seen to have had square spokes, ten or twelve to a wheel of 1–1.2 metres diameter. Sixteen chariots were found in nine cemeteries.

Scholars disagree as to whether the chariots were war machines or only used as ceremonial vehicles. Their dissentions are based on the gauge of the wheels, their distance apart from each other. Egyptian chariots have a gauge of 1.54 to 1.80 metres, while steppe chariots have wheels about 1.50 metres apart. Along with the chariot remains in one cemetery, the graves of two adults, presumably the sacrificed crew, were found. The suggestion has been made that this chariot would have been too narrow for a fighting vehicle. But six other chariots have a gauge as wide as an Egyptian chariot and two Petrovka chariots have gauges of 1.4 to 1.6 metres, large enough for a crew of two. So the question is not settled.

Anthony (2007) favours the idea that the chariots were used for fighting. He cites the discovery of weapons with the chariot crew burials and makes many suppositions based on his desire to present his view. But I feel that, although his arguments are based on suppositions and not hard evidence, he is probably right. If a vehicle can be used in fighting it will be pressed into service.

The bits of the horses have not survived but the spiked cheek pieces made of antler or bone have. The average is four spikes in each. Most of the holes for the bits in them are round but some are square or slots, which to me suggests that they were made of leather rather than rope.

In general there is no visible means of suspension from the bridle for the cheek pieces. It must have been attached to the ends of the bits. But at Krivoe Ozero there was a slot in the cheek pieces above the central hole that could well have been for suspension. This just goes to show that the Sintashta culture people were good enough horse masters to treat each animal according to its particular needs. Radiocarbon dates for Sintashta chariot burials are in general before 2000 BC, while the earliest Near Eastern evidence for chariots pulled by horses and with spoked wheels is in Old Syrian seals of about 1800 BC.

Two seals from the *Karum* of Kanesh, the Assyrian trading colony in Anatolia of about 1900 BC, portray solitary drivers in chariots with four-spoked wheels drawn by two equids of some sort. These do not wear bits, but have nose rings like the Mesopotamian onagers that had pulled the by then obsolete Sumerian battle wagons. So they combined the new technology while retaining part of the old one.

A very pertinent question is how much these steppe advances in equestrianism filtered down south to Mesopotamia. For although the Sumerian and then Semitic people of Mesopotamia were in advance of the herders of the eastern steppe in literacy and possibly the arts of government, the steppe people were better horse managers. The little scenes that are engraved on Mesopotamian cylinder seals provide much information.

An impression of the seal of Abbakalla, the royal animal disburser of King Shu-Sin of Ur of the Third Dynasty of Ur period, dated to 2050 to 2040 BC has, apart from its inscription in Sumerian cuneiform, a representation of a man riding an equid, which we are justified in thinking was a horse and not an onager. That is not to suggest that these late Sumerians habitually rode animals. But it had been tried and the technology had filtered through to some extent.

At that time horse bones appear for the first time on the Iranian Plateau and at the site of Malyan the teeth of one horse show wear to such an extent that it must have been caused by a metal bit. While bits had been in use in the steppe for many years, they were new by 2000 BC in Iran.

Anthony makes a strong case for his contention that the horse first appeared in Mesopotamia during the Third Dynasty of Ur period, coming in from Iran and Anatolia. They appear in Sumerian texts as 'The ass of the mountains'. But he does not suggest that the Third Dynasty of Ur Sumerians used chariots. There are no pictorial representations of them that far south from that time. That was for the future in Mesopotamia.

But chariots were in use in Sintashta and the news of technology applicable to war travels fast, although mounted cavalry was a long way in the future. For riding a horse on only a saddle cloth without a saddle demands much greater slowly acquired skill than driving a chariot and riding a horse was not regarded as a prestige form of transport. Zimri-Lim, King of Mari (1776–1761) during the Old Babylonian period was criticized for riding a horse on a state occasion when a man of his importance should have driven a chariot or have ridden a mule. Clearly the horse was solely for the use of the brutal licentious soldiery.

Rulers and those who want to make a statement about their position have always used historical or obsolete means of transport. Sumerian Early Dynastic rulers used sledges after wheeled vehicles were available. Our Queen rides on state occasions in a horse-drawn carriage although she has numerous motor cars. When Christ entered Jerusalem on an ass it was not a gesture of humility, as many preachers have claimed. An ass, not a horse, was the mount of a king.

There were no wild horses in Bactria, which is modern Afghanistan, c. 2100–1800 BC. The natural equids of the area were onagers. Wild horses had not yet strayed south of what is now central Kazakhstan. There were wagons and carts with solid wheels drawn by Asian zebu cattle. Small funeral wagons with solid wheels and bronze-studded tyres are found buried in royal graves dated to 2100–2000 BC. In the three graves at Gonur there were a funeral wagon, a decapitated foal and a BMAC (Bactria-Margiana Archaeological Complex) style seal from Bactria showing a man riding a galloping equid.

These finds suggest that horses began to appear in Central Asia between 2100 and 2000 BC but were never used for food.

Steppe immigrants from the north brought chariots with them. Bronze bits with looped ends and Sintashta-type cheek pieces were found c. 2100–2000 BC in Petrovka graves at Tugai on the Zeravshan River in southern Central Asia. Bronze straight bar bits appear in graves of 2000–1800 BC in Bactria. These, and bone spiked cheek pieces are steppe-type artefacts that had found their way south of the steppe.

Between 1900 and 1800 BC for the first time a chain of broadly similar cultures extended from the edges of China to the frontiers of Europe. The steppe world was becoming an innovating centre in bronze metallurgy and chariot warfare. The chariot-driving Shang princes of China and the Mycenaean princes of Greece were contemporaries at opposite ends of the ancient world about 1500 BC, thanks to the herders of the Eurasian steppes.

The opening up of Russian archaeology and research since 1991 to non-Russian speakers in the West has vastly increased what we know of the archaeology of the steppe land of Siberia. As far as chariots are concerned this has cast a most illuminating light on the early history of transport in Mesopotamia and has caused a re-evaluation of existing conclusions.

We now know that the technology of horse riding in the steppe was much superior to what it was in the Land of the Two Rivers, and we can no longer be so confident that Mesopotamia was the place of origin of the wheel.

Chapter Four

Sumer

To start with the earliest literate Middle Eastern people, the Sumerians, we note that they spoke a language, Sumerian, which is not related to any other known language. These people were the very early, in fact probably the first, to write it down in a form, that is cuneiform, that we can read. Leaving aside the conditions of settlements in the Indus Valley and Eastern Asia, civilization was first established in the Middle East about 3000 BC, in small city states in southern Mesopotamia; the land between the two rivers of the Tigris and the Euphrates, in what is now Iraq. Sumer was an irrigated desert area in southern Mesopotamia that contained eighteen little towns in the early third millennium BC. It was in the region of the lower Euphrates, inland of the shoreline of the Gulf, which in those days extended up to include the north-west edge of what later became the Iraqi Marshes.

The Sumerian city states lasted for a thousand years before excessive irrigation and resulting salinization of the soil turned the area, then known as Sumer, into desert by about 2000 BC. In that area there were established civilized city states occupied by people called the Sumerians.

The best known of those states were Isin, Larsa, Ur, Lagash and Umma. They were separated from each other by an unoccupied desert that was not irrigated and was known in the Sumerian language as E.DIN, the land outside the city. This is the origin of the later Semitic Hebrew word Eden.

The first recorded war in history – and history consists of the records of societies and very often their wars – was between the Sumerian city states of Lagash and Umma, probably over access to water for irrigation.

Between approximately 5000 and 3750 BC a people dwelled there whose culture was what is now called the Ubaid culture after the type site of al Ubaid where they were discovered by archaeologists. Urbanization had not yet reached Mesopotamia and so there were no wheeled vehicles in their little villages. Horses that could be used to draw warlike vehicles had not yet been domesticated in Mesopotamia and the spoked wheel had yet to be introduced into that region.

The Ubaid culture was followed by the Uruk culture, named after the type site of Uruk, modern Warka, where it was first discovered. The Uruk culture is dated to between 3750 and 3000 BC and excavation at sites where it is found have uncovered the beginnings of urbanization. The little scattered villages had become small towns.

We do not know the ethnic connections of the Uruk people. They may have been Sumerians, but with them literacy had not reached the stage where scholars can identify their language.

Among themselves Mesopotamian archaeologists do not make much use of dates. For they know how uncertain are all the dates before the later days of the great empires in the 1st millennium BC. Instead, they assign excavated objects to a specific culture, to a group of people using the same kind of objects. A culture is often named after its type site, the place where the remains of the objects used by a certain culture, were first discovered. To complicate matters, these sites often have several names. If we know what a settlement was called by its inhabitants in historical times, we generally use that name. Otherwise we use the current local name, or a transliteration of it. Sometimes the local name and the ancient one are both used. A complication that the archaeologist has to be aware of is that in the Middle East place names are often changed for political reasons, usually to remove the recognition of the existence of an ethnic minority from public awareness. But this need not worry the reader of this book.

The cultures in southern Mesopotamia just after 3000 BC come in what is called the Early Dynastic period, which is divided archaeologically into three sections, I to III, which we date to between c. 2900 and 2334 BC. The small settlements had become fortified towns, the capitals of city states such as Ur, Uruk, Isin, Larsa, Lagash and Umma. Civilization to an archaeologist means a stage in settlement where there is division of labour, where people specialize in their occupations and where there is a form of government that has extended to require a bureaucracy for which records have to be kept. At least some of the people are literate.

Writing was developed. This took the form of wedge shapes that were pressed with a stick into wet tablets of clay. The earliest ones discovered are store lists in the form of pictograms; pictures of the items that are listed. Soon these developed into ideograms where, although the signs still represent pictures, yet their form is so complicated that they cannot be identified by the unlearned as the object they represent. Eventually they come to represent syllables that we can read.

Whether these people came into lower Mesopotamia in the Ubaid or the Uruk period or at some other time is unknown to us. But we do know that the peoples called the land Sumer and so they are known to us as the Sumerians.

Before knowledge of the wheel entered Sumer the Sumerians used sledges These were certainly used for dignitaries to take part in processions and probably also to move heavy loads around. Little clay discs with a central hole and sometimes with painted rays on them that might be representations of spokes have been found at various sites in southern Mesopotamia dateable from between the late Neolithic and the Ubaid period but they are not accompanied by model vehicles and are really much too early for spoked wheels in that region and could have been intended for all sorts of purposes. The best academic opinion, with which I agree, is that they are not model wheels. Sledges were perfectly satisfactory as means of moving loads over short distances in regions where the technology of the wheel had not yet been adopted. In any case, the opinion has been advanced that the land outside the Sumerian towns would have been so criss-crossed with irrigation ditches that wheeled transport would have been of doubtful use except for infrequent but essential military purposes.

Illustrations of sledges appear in the Sumerian pictograms found in level IV of the Uruk period at Sumerian sites, just before the full-blown Sumerian culture of the Early Dynastic period. They depict a sledge with upturned runners that curl up at the tip and have on them a square body, probably of openwork, that has a semi-circular or pitched roof.

That they represent sledges is made certain by the discovery, by Leonard Woolley at Ur, of the decomposed remains of a sledge dateable to the Early Dynastic III period (c. 2550–2250 BC) five hundred years after the first Sumerian pictograms. Also, a proto-Elamite stamp seal from south-east of Sumer shows a similar sledge being pulled by an ox. This vehicle has an openwork box-shaped body within which sits a passenger, perhaps a god, while before him a naked driver stands on the runners.

The sledge that Woolley found probably belonged to a queen called Pu-Abi. The woodwork was very decayed but gold lion masks and strips of mosaic of tiny squares and triangles of shell and stone were found with the sledge in straight lines. They have been interpreted as the remains of decorative trim on the bodywork and enabled Woolley to reconstruct the shape of the sledge, which was in accordance with the earlier pictograms. In front of the vehicle were found the bones of an ox that had been sacrificed.

If this was a very archaic vehicle to find in a context of the full flower of Sumerian civilization, one should remember that rulers have often used long obsolete means of transportation on ceremonial occasion in order to emphasize the historical connections of their dynasty.

In the same Uruk IV level as the tablets bearing the pictograms of the sledge were found others that had showed two circles under the runners. These have been interpreted as wheels, the first in Mesopotamia. By the Early Dynastic I period (c. 2750 BC) the early pictograms had become ideograms and a part had started of the long, complicated, and fascinating story of the transliteration and translation of cuneiform by linguists who were able to read the Sumerian language, in which a wheeled vehicle was called GIGIR. To summarize, it would appear that in Mesopotamia the wheel was first adopted at the time of level IV of the Uruk period (c. 3000 BC) when the inhabitants of southern Mesopotamia were Sumerians.

The two actual Sumerian vehicles known to us are the two-wheeled battle-cart, called by some archaeologists the astride chariot or cart, and the four-wheeled battle wagon. From the Early Dynastic period we get the first good evidence for wheeled vehicles in Sumer. A copper model of a light vehicle assigned to the Early Dynastic II period was excavated at Tell Agrab in Sumer. It is drawn by four equids under a yoke, but there the similarity to the four-wheeled battle wagon ends, for it has no space for warriors to stand beside or behind the driver who sits astride the body of the vehicle. Before the driver is an openwork front that is just high enough to save him from being pulled over onto the team by the reins. It has been called by archaeologists an astride cart and it has no clear function in warfare.

Another kind of astride cart is engraved on an Early Dynastic III limestone relief from Ur that shows clearly the type of tripartite solid wooden wheels that are found on battle wagons of the ED III Standard of Ur. It has an animal skin or fleece thrown over the seat, which demonstrates that although there is no driver sitting astride it, it is an astride cart like the Tell Agrab cart and the driver would sit on it as if it were a saddle, although a saddle for a ridden equid had not yet reached Mesopotamia, if indeed it had been invented at all. And it is doubtful if anyone tried to ride a fractious onager. The cart has a quiver lashed to the front that contains javelins like the quivers on the Standard of Ur vehicles and it contains as well a couple of battle axes. The reins are missing, as are the upper parts of the draught animals. But from their feet we can see that they are lions. The assumption must be that the cart is taking part in a cult or political procession.

Also in an ED III context at the Sumerian city of Lagash (modern name Telloh) was found a limestone stele that celebrated the victory of Eanatum, King of Lagash, over Enakale, King of Umma, another Sumerian city state twenty-five miles away. The stele, which has pictorial carvings on each side, has been given the modern name of The Stele of the Vultures because of the way in which on one side the people of Umma are depicted wrapped in a net while vultures fly overhead. If this is not the first war in history, it is certainly the first well recorded one.

Lagash and Umma had been at loggerheads for a long time over water rights.

In this case, according to Eanatum, Enakale had taken possession of some agricultural land between the two cities and the gods had commanded Eanatum to attack Umma. Eanatum won the ensuing war.

On the other side of the stele from the one just described the army of Lagash is shown advancing dressed in helmets and cloaks and clutching spears. In a register below that in front of massed soldiers is a man in a vehicle. At this point the stele is badly broken and much is missing so that we do not see the wheels or the draught animals. But we can assume that the wheels are of the usual Sumerian solid wooden tripartite type. There is certainly only one man in the vehicle, and it is so short that it must be a two-wheeled cart and the driver is sitting in it, not astride it.

Also from Sumer, from grave number PG 779 in the so-called Royal Graves at the city of Ur, is the most famous and most important piece of evidence we have for the wheeled fighting vehicle in the Near East, if not the world, the Standard of Ur. This is an inlaid wooden box, now considered to have been the rectangular sounding box of a stringed musical instrument dated to late ED II or early ED III. On each of its sloping sides, which measure 19 by 8 inches, it bears inlaid scenes in three registers of shell figures on a lapis lazuli background. The whole is affixed to the box with bitumen.

On one side is what is called the Peace Scene, which shows a banquet celebrating a victory. The king, wearing the typical Sumerian fringed skirt, the *kaunakes*, sits in a chair facing a row of courtiers who hold drinking goblets. Behind them is a musician who plays a harp. Behind the musician is a woman singer.

The two registers below show servants bringing cattle, sheep, and fish as food for the feast, while porters carry what is interpreted as spoils from the successful war.

On the opposite side, the War Scene, the top register shows the king, distinguished by his greater stature, who has dismounted from his onagers-driven four wheeled battle wagon. Attendants or soldiers wearing helmets and the typical Sumerian fringed skirt, but not the heavy cloaks that Sumerian troops wore in action, lead naked prisoners up to the king.

In the centre register we see Sumerian heavy infantry drawn up in a rank. They wear helmets and over their *kaunakes* a heavy cloak covered in small circles, which have been variously interpreted. I favour the idea that they are metal discs to ward off blows. For weapons they carry thrusting spears held horizontally pointing forward at waist level. Ahead of the heavy infantry are skirmishers wearing helmet and *kaunakes* but no cloak. They engage the enemy who are naked. Some of the enemy are clean shaven like the Sumerians. Others are bearded. Woolley suggests that these latter are possibly wild men from perhaps the hills of Elam east of Sumer. They would not be Sumerians.

The lowest register is the most important for our purpose and the principal source for our knowledge of Mesopotamian fighting vehicles. It shows four battle wagons, or perhaps the same battle wagon as it picks up speed, charging over fallen enemies. This kind of vehicle with its solid wheels is regarded as a firmly established step in the early development of the wheeled vehicle. For further information on early horse-drawn vehicles scholars in the West must await the publication in a language understood at a sufficiently advanced level outside Russia of the results of excavation in the Ukraine and Siberia. For they show that, with the widespread use of the snaffle horse bit in the Ukraine and Siberia, the Sumerians were not in the lead as far as equid transport was concerned but were actually, with their onagers' nose rings and their vehicles' solid wheels, behind the latest technology. Their battle wagons were in many respects rather archaic obsolete vehicles. But to the defence of the Sumerians it would be only fair to point out that while we have evidence for more advanced technology in wheels and harness in Central Asia, we have no evidence that the people there used their horse-drawn vehicles as part of organized disciplined armies in warfare. Individual disorganized chariot drivers galloping into Europe do not constitute the first chariotry. The Sumerians with their battle wagons do. And this was in spite of their having to do it with the ferociously wild onager rather than the much more tractable horse.

And they were more advanced than the horsemen of Central Asia in that they were literate, which the Ukrainians and Siberians were not.

The body of the long narrow vehicle on the Standard of Ur is of heavy spars with the spaces between them panelled in what is taken to be leather. A painting made some time ago portrays the vehicle as if this high feature at the front end of the vehicle was actually on the side of the battle wagon. But it is now generally accepted that it was actually the front of the wagon and it was placed on the Sumerian inlay as if it were on the side because they could not cope at that time with perspective. In any case the full view of it is most useful to the person wishing to build a replica of the battle wagon because taking it as the front of the battle wagon gives us the internal width of the body. It was long and narrow so that the two men of the crew, driver and warrior, had to stand one behind the other.

On the Standard the warrior on the battle wagon is shown holding a thrusting spear or a throwing javelin. It is probably the latter, or else there would be no need for the quiver full of them lashed to the side of the front. Although a Sumerian text tells us that in his war with Umma the king Eanatum of Lagash was struck by an arrow, there is no pictorial evidence of the use of bows and arrows in Sumerian warfare. Except for this one occasion they must have been restricted substantially to hunting.

The wheels of the battle wagons on the Standard of Ur are shown clearly. They are of the tripartite type with the central plank being a four-sided lozenge shape like the wheels on the astride cart. Ancient pictures of other tripartite solid wooden wheels are known. But in them the central plank has parallel sides, unlike the ones illustrated on the Standard of Ur and on the limestone relief depicting the astride cart from Tell Agrab described above.

Actual wheels have been excavated from Early Dynastic dated sites. The carbonized remains of the wood of two pairs of wheels estimated to date from the ED II period, whose rims were protected with copper nails, were excavated from a tomb in Susa, a city outside Sumer. The three planks from which the wheels are made are fastened together with ligatures thought to be made of leather. There is no sign of nails round the circumference of the wheels on the battle wagons on the Standard of Ur but these are indicated in the painting of a four-wheeled battle wagon on the vase from Khafaje described above. The Susa wheels also displayed other local technical variations that did not occur at Ur, but which indicate that for all that, these solid wheels were archaic and obsolete compared with what was being done in Central Asia.

Yet considerable experimentation in wheel technology was being undertaken in southern Mesopotamia in the early 2nd millennium BC. The Susa wheels had wooden felloes four centimetres thick and it was into these that the nails round the rims were driven. In the earth in which these very decomposed wheels were found eight small copper nails were situated at the centres of the wheels. These might have been bearings to protect the holes for the axle at the centre of the wheels from being worn away, or else they served to fasten hard wood rings that themselves acted as bearings where the wheels turned on the axle. The wheels were 0.85 of a metre in diameter.

After the three Early Dynastic periods of Sumerian history there were several periods when the chronology is not at all certain, with co-existing local dynasties, and we are dependent on our knowledge of the names of major rulers in the area . First came the Dynasty of Agade when from about 2371 to 2230 BC at least the rulers seem to have been Semitic Akkadians. Then there is The Third Dynasty of Ur, a Sumerian renaissance from 2113 to 2006 BC, right at the end of the time when the Sumerian city states flourished for a last time, before the land reverted to desert and Southern

Mesopotamia came under Semitic Babylonian rule. That is followed by the Isin-Larsa period when these two Sumerian cities were rivals for importance until about 1783 BC when the Babylonian Hammurabi conquered Isin and ushered in the Old Babylonian period.

The people of southern Mesopotamia were an inventive lot who were eagerly investigating the technology available to them. But did they know about the latest invention of spoked wheels in Central Asia, or had difficulties in communication kept this from them? Or did they prefer to stick to the old technology of the solid wheel, which their wheelwrights could cope with? Knowing of a new technology that would give them faster vehicles would have been no help if they were restricted to using unreliable onagers instead of more trainable horses and they had no craftsmen who could make spoked wheels. Literacy was confined to scribes whose tasks were to act as bureaucratic clerks or compilers of religious texts and annals glorifying the king's victories in the service of the gods. The technology of wheeled vehicles is a subject that would not appear in writing until the invention of printing three and a half thousand years in the future spread literacy throughout the laity of Europe.

Several Sumerian words for different kinds of equids appear in Akkadian and Third Dynasty of Ur deposits and we can be fairly sure that the horse (*Equus caballus*) was known in Mesopotamia by Akkadian times at the end of the 3rd millennium BC.

Cylinder seal impressions of the early 2nd millennium BC show personnages who are probably deities standing or sitting on vehicles that have two wheels each with four spokes and by that time the inhabitants of Mesopotamia seem to have adopted the spoked wheel found earlier in Central Asia but retained the old inefficient nose ring when the bit was known in the north. The earliest excavated horse bits from the Near East date from after the 15th century BC. Excavated skeletons suggest that the horses stood about thirteen hands high, the size of a modern child's pony.

Chapter Five

Mesopotamia Between the Sumerians and the Assyrians

This period, where dates cannot be so exactly fixed as they can later on, covers Mesopotamia from approximately the 21st to the 12th centuries BC. It extends from the end of the Third Dynasty of Ur and the end of independent Sumerian city states to the rise of Assyrian power in the 12th century BC.

This covers the Old Babylonian Period, the Kassite invasion of Babylonia from the east and their takeover of the country, and the whole of a time when the peoples of Mesopotamia, Syria and Palestine were in violent movement. There were almost continual wars between new peoples looking for lands in which to settle and established powers who were fighting to retain what they had, or old small tribal ruling families were intent on building up their power so that one day they might rule an empire.

Evidence for chariots in the period of Kassite rule of Babylonia of the 17th to the 13th centuries BC is poor as it comes from the engravings on seals or pictures of chariots on Kassite *kudurru* boundary stones where the scale of the drawings is very small. But the vehicles seem to have been light, with six spokes to the wheels, and as far as we can tell they may have had openwork bodies and the axle in the centre of the body. The crew probably numbered two, driver and archer. They had a draught pole from the floor of the body and another support running from the top of the front of the body to the yoke. These look remarkably like the early second millennium Assyrian chariots of the reign of Ninurti-Tukulti-Ashur (c. 1133–1132 BC).

The Kassites probably learned chariot building from the Hurrians who occupied the area between Anatolia and Syria, in the foothills of the Anatolian highlands.

Excavation in 1939 of the archaeological site of Yorgan Tepa, near Kirkuk in northern Iraq, uncovered the remains of the 15th century BC Hurrian city of Nuzi, which was a provincial city of the Hurrian kingdom of Arrapha. In addition to a few small clay models of chariots, the site contained four thousand readable clay tablets in cuneiform Akkadian, which provide us with a full and fascinating view into life in a city in the aftermath of the Sumerian city states and before the rise of the great Mesopotamian empires of Assyria and Babylon. Two historical periods were identified. One was possibly, for dating is not at all exact, from the period of the flourishing of the Sumerian city states when the settlement was called GA.SUR in Sumerian, before the Hurrians moved in. They seem to have done this gradually and peacefully, when what was now a city called Nuzi was established.

The earlier GA.SUR levels produced small clay models of chariots, which had high fronts, underneath which were holes for the wooden axles of the wheels. There was space for only one figure standing on the platform behind the front. These little models closely resembled the many similar ones that are found at many other sites in Iraq. On the basis of that, and illustrations on engraved seals of gods being led in processions on chariots like these, they are considered to be religious votive objects and not illustrative of actual war chariots.

The Hurrian levels at Nuzi produced similar clay models of chariots. But what is special about Nuzi is the cuneiform tablets that give tantalizing snippets of information on aspects of chariots as well as on trade, agriculture and everyday life.

Various types of chariots (*narkabtu* in Akkadian) are mentioned. They include chariots described as 'of the country', 'of the mountains', 'with a shoe', which must be to slow the wheel when descending a steep hill, and a 'swift chariot'. In a list of metal borrowed from the arsenal there is mention of a chariot inlaid with gold. More practical war chariots had protective mail on them, which must have been of lamellar bronze plates. So that gives us some idea of what the war chariots would have been like. The tablets include details on the equipment of chariot crews, their armour, swords and bows, and their arrow quivers, which contained thirty or forty arrows. We also learn that the chariots were painted and the colours included vermillion, and purple or red. A small clay model of a chariot yoke showed that the Nuzi chariots were pulled by two horses. Unfortunately, no illustrations were discovered to show us what a Nuzi war chariot looked like.

On the lighter side I cannot forbear from mentioning that the tablets include the business records of a marriage bureau run by the wife of a palace official who found suitable husbands for girls up from the country. As Hurrian commercial law at that time in a non-money economy did not envisage commercial contracts between the director of the bureau and her clients, she got round this by adopting the girls as her daughters. When they were introduced to a prospective husband, they gave their 'mother' a previously agreed present.

The chariots and armour worn by the Nuzi chariot crews would have been of the same kind as those employed by the Hurrian charioteers of the kingdom of Mitanni, which are described in chapter 11. Two horses pulled a chariot, which is thought to have been light, rather Egyptian in style, with most often four spokes to the light wheels that were oiled to prevent warping in the hot dry climate. Occasionally there were six or eight spokes that spread the shock of impact on the ground. The axle was most often at the rear of the body but occasionally under the centre. The body was open, without panelling between its principal structural members. The curved draught pole sprang from the floor of the vehicle and there was an additional straight solid support that came from the top of the front of the chariot body and joined the draught pole at the yoke that went over the horses' shoulders. Crossed quivers were strapped to the sides of the chariot, one for arrows and the other for the archer's bow when it was not in use. This compound bow was a delicate instrument that had to be protected from extremes of climate. The horses were well protected from enemy arrows by a caparison or blanket of horse hair three centimetres thick that protected the horses' backs from the withers to the loins. Apparently twenty-two minas of horse hair (when a mina weighed 1.25 pounds or 0.571 kilograms), were required to make coverings for a pair of chariot horses. The protective coverings might themselves be covered with leather or lamellar scale armour.

As the chariot offered no protection, the two men of the crew had to wear armour. Two types of lamellar scale armour are described in the Nuzi texts. They do not differ greatly from each other except in the degree of protection they provide and their weight. Both are worn over a wrap-round robe that reaches down to mid-calf length. The heavier one has armour that goes down to mid-calf like the under garment and has wide elbow-length sleeves. On to the presumably leather foundation garment of this armour were stitched 500 large rectangular bronze scales of armour with another 500 smaller scales providing protection for the arms as far as the elbows. The wearer's neck was protected by a high wide bronze collar. A lighter type of armour was similar in style but extended only to the knees and the sleeves covered the arms only half way to the elbows. With both types of armour the crew men wore a helmet of two rows of lamellar scales from which hung a feather or fabric plume. We have to presume that the driver was as well protected as the archer.

The Hurrians are credited with inventing this type of scale armour about the 17th century BC and it was adopted by many other armies in the Middle East, even if not in such a whole enveloping form.

Nuzi is a particularly interesting Bronze Age Middle Eastern site because the very large number of cuneiform economic tablets found there gives us a unique insight into how the exchange of goods and services worked in a non-money economy. For money, a portable means of exchange, was a classical Greek invention and considerably later than the Hurrian occupation of Nuzi. At Nuzi a man who produced goods, be they baskets or arrows, took a selection of what he produced along to the market outside the city gate and swapped them for what he needed in the way of goods and services. Of course the tablets are not a comprehensive analysis of the Nuzi economsy. They are just the records of those exchanges that were important enough to necessitate a written record. They do not mention how a person, like for example a chariot driver, who did not produce an immediate object or service for an individual was rewarded for his labours. But we have enough records from Egypt and Assyria to show that a man who performed some such service for the state was provided with land for himself and fodder for his horses and so could look after himself.

Immensely informative as the Nuzi tablets are about the life of the Hurrians in the 2nd millennium BC, yet they do not answer the question of what were the linguistic affiliations of the Hurrians before they entered Mesopotamia or tell us what their chariots were like. For we do not know whether their design tended towards that of the light Egyptian pattern or the heavier chariots employed by the later Hittites and Assyrians.

Chapter Six

The Early Assyrian Empire

In the early 2nd millennium BC after the homeland of the Sumerian city states had reverted to desert, central and northern Mesopotamia were occupied by a Semitic language speaking people, the Assyrians. Their capital city was Ashur, with Nineveh another important city. To the south of them, occupying territory that had included the northern border of the area of the by now deserted city states of Sumer, was another closely related Semitic people, the Babylonians, with their capital in the city of Babylon, literally The Gate of God. Both countries followed similar religions with a number of gods who oversaw areas of human endeavour, although not as closely as the classical Greek and Roman gods. Each state had a principal god who was the national deity. With the Assyrians it was Ashur and Babylonia's national god was Marduk.

We do not know where the Assyrians and Babylonians came from, although it was possibly Arabia. The stages by which they occupied what became Assyria and Babylonia are not recorded by history. Little is known about the military and political history of Babylonia before the fall of Assyria, except where it impinged on Assyria. But a great deal is known about Assyria before the collapse of its empire in 612 BC. The successive Assyrian wars were waged because of the religious duty imposed on each Assyrian king by the god Ashur. This was that each year after the harvest the king should lead his army on a campaign to increase the power and prestige of the god Ashur and the wealth of his people, the Assyrians. So each year the Assyrian kings ordered a written account of its campaign to be written for the god Ashur in the form of the king's annals. In approximately every second year they raided the prosperous city states of Syria and Palestine to extract tribute from them. In the intervening years they fought defensive campaigns against the incursions of the people of the Zagros Mountains to the east and of the Babylonians to the south. We know that after each three-month campaign the soldiers of the Assyrian army were returned to their villages and resumed their occupations as peasant farmers.

The language of the Assyrians was Assyrian, a member of the Semitic language family. It is reasonably closely related to Hebrew and very similar to Babylonian. All eight dialects of this Semitic language are now extinct, although well-known to modern Assyriologist linguists who call them collectively Akkadian. Official records of it, including the kings' annals and official communications, are written in cuneiform in an official civil service dialect known to modern linguists as Standard Babylonian.

Three Semitic languages are spoken today: Arabic, Modern Hebrew and Aramaic. The last one is still spoken in parts of Northern Iraq and Syria.

Even today in the Middle East the local dialect of the generally spoken language can change every two kilometres, so it is not surprising that there was such a plethora of dialects over a long time scale in antiquity.

Akkadian is not an impossibly difficult language to learn for someone who is familiar with how Semitic languages are constructed, but the system of writing, cuneiform, is fiendishly complicated. Akkadian cuneiform, literally wedge-shaped, is a syllabic script, not an alphabet. It was derived from the cuneiform script of the unrelated Sumerian language. Although principally impressed on clay tablets, it could also be carved on stone. Other languages than Akkadian were also written in variant forms of cuneiform, but for a study of official Assyrian records we are concerned only with Akkadian in the Standard Babylonian dialect. In syllabic Akkadian cuneiform there were a vast number of signs to portray what in alphabetic English are two or three-letter syllables. Old Babylonian of about 1800 BC has around a hundred and fifty different cuneiform signs and by the time of Standard Babylonian in the 1st millennium BC the number had risen to some six hundred and fifty. There is no recognizable system by which the signs are related to each other. To complicate matters slightly, certain commonly used words such as 'man' and 'horse' are written in an Assyrian cuneiform version of a Sumerian ideogram where one sign represents a whole word. But in reading, it was pronounced in Babylonian. For example, 'horse' is written in the Sumerian ideograms ANŠE.KU.RA (Ass of the mountains) but pronounced in Standard Babylonian as Sise. We know how it was pronounced because of the language's close resemblance to Hebrew. Words are run together in cuneiform without a break and the only punctuation is the paragraph. But in antiquity only professional scribes could read the cuneiform writing and as it was a hereditary occupation they had a vested interest in keeping it as complicated as possible. To this day being able to read Akkadian cuneiform is a specialist branch of Mesopotamian archaeology.

Each year the Assyrian king had his annals carved in cuneiform on stone or impressed in clay, starting with a bombastic account of the success of that year's campaign. So we know a lot about the Assyrians' military operations.

We are accustomed nowadays to governments proclaiming themselves to be more humane than their military actions subsequently show them to be. The Assyrians went in the opposite direction and advertised themselves in the annals as having been exceptionally brutal. With that they had the intention of frightening prospective enemies into being so frightened of the Assyrian army's approach that they were prepared to meet Assyrian demands before they arrived. The Assyrians' reputation for 'deliberate frightfulness' and carefully directed atrocity would precede their army and so they would encounter minimal resistance. They would be able to impose on a wealthy western Syrian city what demands for tribute in the form of gold, silver, moveable goods, livestock and captives that they needed to finance the Assyrian state and its annual campaigns.

Leaving aside the control of natural resources, there are two possible reasons for conquering a country. One is to acquire land in which your own people may settle. Another is to acquire wealth. If it is the land you want, the indigenous population is surplus to requirements and may be massacred. If it is wealth you want, then you allow the local people to remain and work the land for your benefit and pay tribute to you instead of taxes to a government chosen by themselves. The latter is what the Assyrians wanted. In advance of their army's arrival at a city which they wanted to exploit, the Assyrians announced how much tribute they would accept. If this was refused, the amount was increased. If the leaders of the local community still held out, the city was besieged, an operation at which the Assyrians were highly skilled and which caused them fewer casualties than open battle. When the siege was successful the Assyrians collected what wealth they could carry away and imposed an annual tribute.

The leaders of the defeated city, the elders of the town council that had rejected the Assyrian demands, were then executed with every possible barbarity, like impaling and flaying alive, as a warning to others. Their deaths were illustrated dramatically in the stone reliefs that lined the walls of rooms in Assyrian palaces showing the atrocities that were imposed on those who opposed the might of Assyria. When a foreign tributary king or ambassador was summoned to appear before the king of Assyria, he was conducted through a series of rooms portraying the invincible military might of Assyria and the fates of those who opposed an Assyrian invasion. By the time he reached the throne room he was in no condition to stand out for better terms.

What is now called the Assyrian Empire was not an empire as we now understand the word. That is to say, it was not an expanse of foreign terrain that was occupied by a major power and governed by it according to its own system of government or one that suited it and followed its own moral standards. Rather, it was a sphere of influence into which in the east, south and north, it would not let any other political entity encroach. Any such incursions would be repulsed by military attacks that did not aim at permanent occupation of the intruder.

After the Assyrian king had imposed an annual tribute on each city that had been successfully attacked, leaving the local rulers in power, he returned home with his army. The kings' annals suggest that if a city held out successfully against the Assyrian siege, the Assyrians cut their losses and moved on to the next wealthy city.

A local ruler of an area that the Assyrians controlled had, in addition to paying an annual tribute, to agree to entering into a loyalty treaty. Breaking that would incur the wrath of the Assyrian gods, and bring about an invasion by the Assyrian army. This gave the Assyrians a political and religious god-given justification for their future actions. They were not the only country that has always sought a moral reason for going to war.

If the new local ruler who was installed proved untrustworthy, then an Assyrian governor would be appointed. Gradually in the course of the expansion of the Assyrian sphere of influence more and more of the vassal states were formally taken into the status of parts of Assyria.

If tribute had been levied on a city and in subsequent years it was not paid while the Assyrian army was absent on campaign elsewhere, then at the first opportunity the Assyrians returned to the defaulter, imposed a heavier tribute, and if they judged it appropriate, in extreme cases executed the local leaders with every possible cruelty and appointed new leaders there. There seems to have been always those who welcomed the chance of a little local power, as there were those who believed, against the evidence, that the Assyrians would not return if they stopped paying tribute.

The Assyrian war aims were initially security and the protection of trade routes. Then they became the way of financing the military power necessary to achieve the primary aims and the maintenance of a high standard of living with, as secondary aims, the promotion of national prestige. In the last days of the Assyrian Empire foremost was the protection of the state against minor powers such as the Medes and the Scythians who had learned the advantages of coalition in order to equal and exceed Assyrian numbers.

The trade routes that the Assyrians protected from the intruders from the Iranian plateau were those that produced silver, lead, and tin, used to make bronze for weapons. In the late empire, iron, a strategic material that was ten times more costly than gold, was a material whose supply had to be controlled. In the last century of the empire, from the time of Sargon onwards, the Assyrians were increasingly fighting defensive campaigns against intrusive foreigners whose coalitions were matching if not exceeding in size Assyrian numbers.

We do not know the battle tactics used by the Assyrians, although they seem to have had a preference for sieges. Certainly the Assyrian battle reliefs show sieges, with chariots and cavalry pursuing fleeing enemies after the encounter was already won, but do not illustrate when or how the Assyrian shock forces were employed tactically.

The horse (*Equus caballus*) was domesticated and tamed by the beginning of the 2nd millennium and in Mesopotamia had replaced the less tractable half-ass, the Mesopotamian onager (*Equus hemionus hemippus*) that the Sumerians had used. With the development of the saddle and the stirrup still in the future, riders during the Assyrian Empire could have had only a very insecure seat on a blanket or an animal skin. So horses before the reign of Ashurnaṣirpal II (883–859 BC) were confined as far as their employment in warfare was concerned to being used as draught animals to pull chariots.

A problem that we have in analysing and attempting to construct a chronology of Assyrian harness is that we only have reliefs that depict horses from the reigns of a few selected important kings, and their reigns are sometimes separated from each other by long intervals when insignificant and inactive rulers occupied the throne.

The first chariot scene that we have is from the reign of Ninurta-Tukulti-Ashur (c.1133–1132 BC). The chariot body is of openwork and the small wheels that are at the rear of the body have six spokes. Then we have a gap until the time of Ashurnaṣirpal I (c. 1050–1032 BC). Our next source of evidence after that is on a tiled orthostat of Tukulti-Ninurta II (890–884 BC) from Ashur.

Early Assyrian chariots were very small, with wheels no more than possibly two feet (61 cm) in diameter. They had six spokes and the axle was underneath the rear of the body. That is an important feature considered by modern students of chariots to have given greater stability to the vehicle in a turn than when the axle was under the centre of the body. The rear axle certainly seems to have been a slightly later modification, which was presumably the result of experience. But the difference in stability between the two axle positions has never been tested experimentally.

The chariot of the time of Ninurta-Tukulti-Ashur would be drawn by two horses side by side under a yoke, although only one is shown in the picture we have of it. It is common for Assyrian chariots in pictures of them to be drawn by an odd number of horses, fewer than could possibly be used in practice. This is now often attributed to a desire to avoid cluttering the picture. The animals were small, probably no more than thirteen hands high, the size of a modern child's pony. This was the natural size of horses before selective breeding produced larger ones. No details are visible of the harness of this very early Assyrian chariot. This is followed by the first representation on a detailed relief. It comes from the reign of Ashurnaṣirpal II (883–858 BC) and is followed by reliefs and the embossed strip cartoons of his campaigns on the bronze clad gates of his country palace at Balawat. This was a few kilometres north-east of Nimrud and was occupied by Ashurnaṣirpal's son Shalmaneser III (858–824 BC). By this time Assyria was becoming so important, or at least sufficiently self-important, for successive kings to line the walls of their palaces with reliefs of their prowess in hunting and war.

The next reliefs are those of Tiglath-pileser III (824–745 BC), followed by those of Sargon II (722–705 BC), Sennacherib (704–681 BC), and the last important Assyrian king, Ashurbanipal (665–626 BC).

Those who have made a study of the Assyrian reliefs have had to come to the regretful conclusion that while the principal features of them may have been drawn or carved by skilled artists who were working from life, much of the background and repeated features may have been filled in by copyists working on a production line basis. So some of the repeated features cannot be trusted to belong accurately to the horse of the period depicted.

Assyrian chariot and cavalry horses after the reign of Tukulti-Ninurta II wear a rounded pad retained by a net on their brows. This must have been to absorb blows to the head. Horses are particularly sensitive to blows, even very light ones, on the top of the head. There is an old cavalry maxim, 'In a scrimmage never mind the man. Go for the horse's head.' It is the most effective place to aim for and also the nearest one. And if you manage to sever the head piece of the bridle, the rider will have lost all control.

From the reign of Tiglath-pileser III onwards horses drawing chariots carrying the king in procession have three superimposed tassels on the top of their heads in addition to the brow pad. This feature appears to be restricted to kings' horses. From the reign of Tiglath-pileser III onwards to the time of Ashurbanipal chariot horses being used in war had small forward curved crests similar to those found on men's helmets fastened to the head pieces of their bridles. While the brow pads would have absorbed blows to the front of the head above the eyes, these crests would have absorbed or deflected cuts to the top of the head.

But the tactical use of both chariots and ridden cavalry horses did not change much until horses disappeared from the battlefield in the early twentieth century of the current era. The advantage of a chariot and a cavalry horse over a foot soldier has always been their speed and the fear of being crushed that they engender in the infantryman. They are vulnerable if brought to a standstill, so they must keep moving. A horse, having an acute sense of danger and regard for its own safety, will not charge through unbroken infantry that stands with weapons levelled to receive it. So the chariot charge should only be mounted when as the result of infantry or archers' attacks the enemy morale is shaken and he is on the verge of

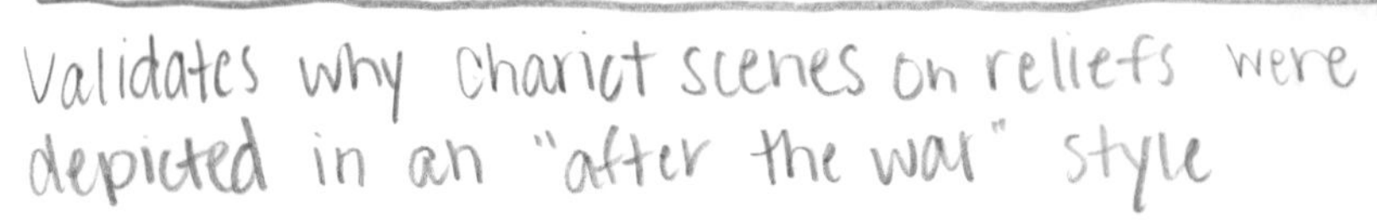

retreating in disorder. Always remember that when the enemy is within range of your weapons, you are within range of his.

It was part of every king's public image to be portrayed out hunting wild bulls or lions with his bow from a chariot.

Without natural frontiers and occupying land that was more fertile than the hills to the east and the north and the desert to the west, the Assyrians were very susceptible to invasions by the hill dwellers who were in search of a richer place in which to settle. This could have been a contributing factor to their concentration on building up their military power. The annals of very early Assyrian kings, however informative on Assyrian political and military policy, tell us nothing useful on the development of chariotry employed in their furtherance. For that we must go forward to the 11th century BC, when the Assyrians started leaving illustrations of their military equipment.

A reasonably clear view of an early royal chariot comes from a picture on a stele reputed by Madhloom to be from the reign of Ashurnaṣirpal I who reigned from 1050 to 1032 BC. It is carved on a stone stele known to archaeologists as The White Obelisk, which was found at the archaeological site of Kuyunjik in which is the Assyrian city of Nineveh. It is approximately contemporary with the chariot already mentioned above that is attributed to the reign of the previous king, Ninurta-Tukulti-Ashur. Although there is more detail in the picture on the obelisk than on the earlier seal impression with two crossed quivers on the side of the vehicle, the chariots are similar. The horses' harness can be seen clearly. Traction is through a breast band, a method that predates the horse collar and can still be seen in use in the Balkans.

With the 1st millennium BC, pictures of Assyrian chariots are much more numerous and provide greatly increased information. If up to that date there are gaps between descriptions of the chariots of Assyrian kings, that is not because the intervening kings are unknown. The annals of many of them have been discovered. But we do not have illustrations of chariots that can be ascribed to their reigns.

Madhloom illustrates and describes chariots of the reign of Ashurnaṣirpal II (883–859 BC). The chariots shown on the reliefs of that reign are interpreted as having semi-circular rounded fronts. This was probably made possible by the chariot crew consisting of no more than three men including the driver. The chariot had become much more robust in the two centuries since Ashurnaṣirpal I. The body of the vehicle was now panelled and reached to just below waist height. Arrow quivers were attached crosswise on the side of the body. Sometimes they contained a bow or axe and sometimes a javelin. The open rear of the body was closed by a curved shield that was positioned across the aperture.

The chariot was still drawn by two horses, although sometimes a third one is portrayed and occasionally the driver is seen to hold three sets of reins. This is a conundrum that has often been debated. Did the Assyrians ever have three horse chariots? A suggestion has been put forward that the third horse was a spare that could be yoked in if one of the others became a casualty. I am not impressed by that solution. A third horse beyond the end of the yoke would make an unbalanced team, which it would be difficult to keep straight when moving. Replacing an injured horse with another excited one in action is not something any horse handler would wish to attempt, as anyone who has had to back an excited horse into the shafts of a wagon will agree. I think there were just two horses and the third one was added for artistic balance in the reliefs. The representation of bodies underneath the chariots shows that they are being used in war.

A feature that probably started with the connecting pole from the top of the chariot body to the yoke in earlier vehicles had now been developed into a kind of springy shock absorber. The upper pole from the top of the chariot body to the yoke had given way to a lentoid construction with upper and lower curved edges. It is now generally interpreted that these are two thin springy members held in position and restrained by bindings that are covered in decorated fabric. As the chariot bounces up and down the springs flex and absorb the shock that might otherwise shear the joint of the supporting pole with the main draught pole that takes the pull from the yoke to the body. The wheels still have six spokes.

As this lentoid shock absorber was an early feature on Assyrian chariots, this is a convenient place to make this account easier to handle by starting a new chapter that deals with what I have decided to call the Later Assyrian Empire, which covers its last 270 years. For Assyria was at the peak of its power from the reign of Ashurnaṣirpal II onwards, when chariots and mounted horse archers were employed together in the army. There was no definite time when the changeover was made from chariots to cavalry. The two carried on in service together for some time, although I have given my reasons for thinking that cavalry had become the more useful arm for the Assyrians by the 7th century BC fighting in mountainous terrain.

Chapter Seven

The Later Assyrian Empire

It was in the reign of Ashurnaṣirpal II (883–858 BC) that the reliefs first show Assyrian mounted horse archers. So this is a convenient point at which to take up the story of the last 246 years of the Assyrian Empire in which they used both chariotry and cavalry together in their campaigns.

The chariots of the reign of Ashurnaṣirpal II's son Shalmaneser III (858–824 BC) differ little from those of his father's time except in minor details such as the positioning of the quivers on the side of the body and in the horses no longer being covered with a decorated blanket. They continued to be drawn by two horses and the crew consisted of two men in addition to the driver. One of these might be the king out hunting or in battle. In both of those activities he would be shooting with a bow and arrow, leaving them aside only for a procession.

After Shalmaneser III we have a gap of seventy-nine years before we come to the next king whose chariots are illustrated on Assyrian reliefs. He was Tiglath-pileser III, who reigned between 745 and 727 BC. He had two different types of chariots, which can be seen in reliefs from Nimrud. On one, possibly an earlier type, the elliptical device connecting the top of the body to the yoke was still employed. On the other it had disappeared as it had on later vehicles and its place was taken only by a thin connecting strut, which might have been of metal, rope or leather, and which connected the top of the body to the yoke. It would still have acted as a secondary support for the yoke, as it did on later vehicles.

Otherwise there was little difference between the types, both of which display new features. Instead of six spokes the wheels now had eight, which until now had been a feature found on foreign Syrian chariots. We know that with Ashurnaṣirpal II the wheels still had a thin inner rim into which the spokes were sunk and a much thicker outer rim. But from now on the two rims were joined by two opposed pairs of slightly wedge-shaped features that overlapped them. They must surely have been metal brackets designed to hold them together more firmly.

The crew continued to be two men in addition to the driver, but the body of the vehicle displayed differences from those of earlier chariots. Instead of quivers on the sides of the body there were upright quivers on the front corners, which suggest a flat front to the chariot body rather than a curved one. At the rear of the sides these rose up in small semi-circles, which survive in later vehicles. Although useful for a classification of chariot types, these features did not significantly alter the performance of the chariot as a war vehicle. But the general impression gained is that from the time of Tiglath-pileser III onwards the Assyrian chariot, although no larger than earlier models, became a much more robust vehicle.

Shortly before the accession of Sargon II (721–505 BC) to the throne of Assyria two political movements occurred that had profound consequences. Egypt, which until then had been following its historical policy of not involving itself in political affairs beyond its borders, started to take an interest in Palestine. And the Elamites who occupied the mountainous region east of Assyria started to interest themselves in the affairs of Babylonia to the south of Assyria. Tiglath-pileser's advance onto the Iranian plateau had cut across the only trade route open to the Elamites, while his conquest of Phoenicia had cut Egypt off from the wealth of northern Syria.

So the Elamites and the Egyptians joined the Urartians in becoming Assyria's enemies. None of them were yet strong enough to attack Assyria openly alone, so they fostered local revolts against Assyria and assisted the Aramaean sheikhs of Babylonia in their opposition to the Assyrians.

This was the start of a long Assyrian struggle against rebellions by rulers of states in areas which had been previously quiescent and favourable to them. Sargon claimed to have won a victory in 720 BC against a Babylonian/Elamite coalition. But a Babylonian text gives a different version, that the Elamites defeated the Assyrians. Meanwhile in the west the Egyptians fomented trouble for the Assyrians during their efforts to subdue a coalition of Syrian kingdoms.

The Assyrian annals, while accurate enough in saying where the Assyrians campaigned at any time, are not at all reliable in claiming victories. For they are religious/political documents that tell the god Ashur each year that the king has fulfilled his duties as a war leader on behalf of the god.

With Sargon II, Assyria was a major regional power with Nimrud (Assyrian *Kalhu*), the military capital. It contained a palace built by Ashurnaṣirpal II, which Sargon restored. But in 717 BC he had the foundations of a new giant palace *Dûr-Sharrukin* (Sargon's fortress) built for himself near the modern village of Khorsabad north-east of Nineveh. The interiors of its chambers were decorated with the usual reliefs portraying Assyrian success in warfare and the fate of those who resisted it. The surrounding new town measured one and a half kilometres each way. Assyria was in the ascendant and the existing pattern of campaigns that extracted tribute continued with yet greater success.

The chariots of this reign differ little from the earlier ones of Tiglath-pileser III, which do not have the elliptical shock absorber. They continued to be drawn by two horses, although a third horse is often portrayed.

With Sennacherib, the king who succeeded Sargon II in 704 BC and reigned till 681 BC, Assyria was at its zenith and chariots became bigger. They appear, as best as can be deduced from side-on views, to have now a rectangular floor plan. The semi-circular front has gone. The wheels now have eight spokes and are of slightly larger diameter than earlier ones. In most cases the tyres of the wheels are studded with nails. Although nail studded tyres had been known for a very long time, they became the norm with Sennacherib. This could be a modification to cope with the rougher and more hilly country in which the Assyrians were now operating. With Ashurbanipal the wheels in some cases reached in diameter higher than the height of a man. As chariots had now given way to cavalry as the effective mounted arm, I think that this increase in the size of chariot wheels was for national prestige rather than for any improvement in performance that the increased diameter might bring. The crew of a chariot, including the driver, which had occasionally been increased to four under Sargon II, now became four as the standard.

The chariots of Sennacherib and Ashurbanipal were pulled by four horses, although teams of two horses were still occasionally used in the time of the earlier king. The augmented team would not increase the speed of the vehicle, but it would be a distinct advantage in getting started again if the chariot were to have to cross very rough or steep ground where it could be brought to a halt. For the Assyrians had to cross ever steeper ground while traversing the ranges of mountains that had to be crossed during Assyrian campaigns against Urartu in Anatolia. The augmented team of horses would increase the distance that the chariot could be pulled without the team having to stop for a rest Remarkably little pull, measured in pounds, is needed by the team of animals to keep a wheeled vehicle going once it has started. And a horse's ability to absorb and use oxygen is sixty times better than that of a man. The limiting factor, it would seem to me, would be the horses' sure footedness and their ability to pick their way over uneven ground. Some horses are good at this and others are not. One must assume that chariot horses were selected for this ability. Another would be the vehicle's durability. Even on a flat plain an unsprung vehicle takes a terrible beating.

During the seventy-seven years between the reigns of Sennacherib and Ashurbanipal Assyria was at its peak and the campaigns of these kings and of Esarhaddon immediately before Ashurbanipal are recorded in their annals as wars of conquest. But if one looks at who their enemies were and the subsequent histories of those peoples, the Urartians, the Elamites, and the Medes, the wars that the Assyrians preferred to record as wars of conquest appear to be really successful counter-attacks. The Elamites, with their capital at Susa, had occupied the mountainous territory east of Assyria since the 4th millennium BC and by the 1st millennium BC were a people in an expansionist mood. Their ethnic origin is unknown and their language is not related to any other known tongue. Again it was written in a variant form of Akkadian cuneiform.

The Medes were an Iranian people, skilled horsemen, who had arrived in the area more recently than the Elamites and were also pushing west into the more fertile plain between the Tigris and the Euphrates.

Assyrian trade routes to the east passed through Elam. South of Assyria was Babylonia and further to the south of that were the Chaldeans, a Semitic-speaking people who had not all that long before immigrated from Arabia. They were pushing north and to achieve that were allying themselves to the Babylonians and urging them to move north against Assyria. In Anatolia, north-west of Assyria, was the kingdom of Urartu. The Urartians were another people who spoke a language that had no connection to any other known language. Unlike the Elamites and the Medes they were not actively moving against Assyria, although they occupied territory through which Assyrian trade routes passed. The geography of that area consists of precipitous mountain ranges running east and west and separated from each other by wide flat valleys. Cavalry would find it much easier to cross the mountains than chariotry would.

So the strategic function of the Assyrian army was no longer to engage in tribute seeking raids on settled but poorly defended petty states in the Syrian plain but to drive back expansionist mountaineers and keep Assyrian trade routes open in the east and the west where, apart from the economic threat of Urartu, New Kingdom Egypt was exerting itself to control trade in Palestine. If Assyria's tactical methods were aggressive, its strategy was defensive; always a dangerous changeover point for a major power.

The Assyrian Empire was still powerful, prosperous, and internally at peace. Its sphere of influence stretched from the Mediterranean to the Gulf and it had a powerful and disciplined army. But it had to fight continual defensive wars to keep itself secure with an army the majority of whose men were Aramaean mercenaries captured from other armies rather than native Assyrians.

We know that the Assyrian army was composed of infantrymen who included archers, slingers and spearmen. Others were chariot drivers, archers who formed the crews of chariots, mounted archers, engineers, and drivers of baggage carts. But we do not know how it was divided between these categories, nor the numbers involved in each. We have a few accounts of its total size. Ashurnaṣirpal II once said he had 50,000 men. At the battle of Qarqar, when Shalmaneser III entered Syria in 853 BC, he had 120,000 men to face the 62,900 infantrymen, 1900 horsemen, 3900 chariots and 1000 camels of the opposing coalition. Enemy casualties included 14,000 dead,

After 639 BC the annals of Ashurbanipal come to an end, although he lived for another twelve years until 627 BC. The period seems to have been characterized by onslaughts on Assyria by first the Medes and then the Scythians, a highly mobile nomadic people who came originally from the Caucasus and for the past fifty years had been encroaching on Assyria. Assyria was ruled by two more insignificant kings after Ashurbanipal until in 612 BC. Nineveh along with the other major cities of Ashur and Nimrud and other towns were taken by a coalition of the Medes and the Babylonians in an assault in which the Scythians possibly had a hand. The Assyrian Empire was at an end.

Riding lagged behind driving as a means of using horses as a method of transport. In the early 2nd millennium BC when men rode horses they sat on the animals' rumps as people do in the Aegean to this day when riding a donkey. This is the only reasonable way of sitting on a donkey as it is built differently from a horse as it has no withers. These are the ridge formed by the tops of a horse's shoulder blades which prevent the rider from sliding forwards off the animal. By the end of the 2nd millennium BC riders in Mesopotamia had adopted the forward seat immediately behind the withers where the ribs are strongest, the one used to this day. Control of both riding and driving horses was through a bit in their mouths. When the mouth piece was made of metal and not of rope or leather this was a jointed snaffle, which is still the commonest general riding bit today, and it is not a particularly strong bit. It is nothing like as fierce as the European mediaeval or renaissance bit, or the bits used in the Middle East today, which press on the roof of the horse's mouth.

There was little difference between Assyrian driving and riding bridles, except in their outward appearance. The important difference between the saddlery of driving and riding horses was in the piece of material, either of cloth or an animal's skin on which the rider sat. This was doubled over from front to back and there must have been a roller, a tied or laced band hidden from view inside the forward fold, which held it in position. It gives a most insecure seat, but was the best there was before the saddle was invented some two hundred years after the demise of the Assyrian Empire.

Armour made of overlapping bronze plates appeared in the Assyrian army after the reforms of Tiglath-pileser and sleeveless armoured vests were widely issued to the end of the Empire.

Armour made of overlapping bronze scales was worn by Sennacherib's horsemen, spearmen and archers. This took the form of a waist length corselet of overlapping lamellar plates with short sleeves. On their heads the men wore pointed metal helmets and they were shod in calf-length boots. Above these, covering their knees, were what could have been knitted or armoured stockings that extended up to mid-thigh.

In typical Assyrian style the reliefs show the cavalry pursuing a fleeing enemy, which is a correct use of horsemen. For cavalry has no place in siege warfare. But up to the time of Ashurbanipal the king is still portrayed standing majestically in his chariot or shooting arrows from it in a lion hunt or in a battle until just before the end of the Assyrian Empire.

Unlike the Mitannians the Assyrians did not have armour for cavalry or chariot horses. That came into general use late in the Roman Empire with the appearance of the cataphracts of the Persian and Byzantine cavalry.

Cavalry, men riding horses in warfare, first appears in the Assyrian army in the reign of Ashurnaṣirpal II. The cavalry were horse archers and they operated in pairs, both men being mounted. One was an archer who loosed his arrows when his horse was stationary and the other was a horse handler who stood on the ground at the head of the mounted archer's horse and held its reins.

Horsemanship improved after the reign of Shalmaneser III. The repoussé scenes on the gates of the palace at Balawat show some riders sitting in the forward seat while others still use the 'donkey riding' rear seat on the animal's haunches. The point should be made here that the term 'forward seat' as it is used here has no connection with the term as it is used with reference to modern equestrian jumping.

The Greeks, good horsemen, changed over from the rear seat to the forward one in the middle of the 7th century BC, not that long after the time of Shalmaneser III.

It is noticeable in the reliefs that the Assyrian cavalrymen held the reins not only in their left hands, which is required of all cavalrymen, but low down alongside the horse's neck. I can suggest a reason for that. Horses, like people, are not bilaterally symmetrical. There is a tendency for right-handed people, the majority, to sit turned very slightly to the right. This is so slight that it is not perceptible to the eye, but it is to the horse, who learns to accommodate to it by turning slightly to the right. This low left grip of the Assyrian riders could be a conscious effort to keep the horse bent very slightly to the left, and therefore straight.

In addition, when a horse is ridden straight at a stationary object, as an Assyrian mounted javelin man would, the horse's natural response is to edge away to the left so as not to hurt himself by colliding with the obstacle. The rider's desire is to pass by the enemy within reach of his weapon. But the horse does not think of that. He looks after his own perceived interests. So the rider has to push inwards with his left leg to keep the horse from edging out to the left. This is best done by rebalancing the horse by turning his head slightly left, away from the direction in which you are pushing it. Remember that the reins only change the direction in which the horse's head is pointing, not the direction in which his body is directed. I agree that this is modern riding, or at least a method invented two hundred and fifty years ago. But the human body and the horse's have not changed since the Assyrians.

If the horse is trained to this and does it correctly he will travel forward and edge to the right, towards the target. It is so embarrassing to reach the target and make a thrust at empty air because the target is too far away.

After Shalmaneser III archers no longer required a horse holder and could loose their arrows from a moving horse, dropping the reins as they did so. So they had the mobility of proper cavalry in skirmishing and harassing and then they could circle away out of danger. The reliefs show the horse archers shooting at a gallop.

By the reign of Sennacherib the mounted archers were joined by cavalrymen holding javelins. These were gripped overhand. Couched lances whose butts were tucked under the armpit had to wait for the invention of solid saddles with stirrups that would absorb the shock of the impact.

As early as the reign of Ashurnaṣirpal II in the 9th century BC. Assyrian chariot horses wore caparisons, cloths over their backs that gave some protection from arrows and javelins, and by the reign of Sennacherib in the 8th century these were common. For although there was no general provision of a standard pattern, yet the general move was towards increased protection of the horses. But even though by the time of Ashurbanipal the Assyrians were at their peak, yet the portents of their coming decline and fall can be detected by the historian in intimations of a shortage of horses as well as of men. One sign was the appearance of cavalrymen carrying a bow and also a javelin. But one must bear in mind that the organizers of armies have always loaded the front-line troops with extra weapons and equipment to guard against unforeseen possibilities.

One highly relevant intimation appears in a letter to which Ashurbanipal replies. A local commander or official had asked that prisoners be drafted as charioteers, messengers, and cavalry. The king refuses the request, saying that those of the prisoners condemned to death must die just the same. This sounds like the eternal civil service clash of interests between the man on the spot and central government that has grand strategy in mind.

Richard Barnett's study of Assyrian reliefs shows that Assyrian cavalry horse bridles do not fit as clearly into types as we might have hoped. For although many if not most of the horse archers' mounts have only the brow pad and no crest on their heads, others do have it. This could be because of sloppiness among the artists who were working on a 'production line' basis and in drawing and carving horses' heads worked from one master drawing and not from life. Or else it could be because, as in every army, there might be a regulation piece of equipment but this was not always available and before action the soldier grabbed whatever was to hand. But certain trends can be detected and it should be appreciated that these are only trends.

In the early part of the Assyrian Empire mounted archer cavalrymen rode horses that had a brow pad over the forehead but no crest on top of the head. Later on, from the time of Ashurnaṣirpal II and certainly by the reign of Ashurbanipal, the brow pad seems to have been put aside in favour of a small version of the curved crest that soldiers wore on their helmets. The king's horses retained their customary high tassels. Presumably the monarch was awarded better protection than the ordinary soldiers. But his horses' tack and chariot harness had always been different from that of his subjects. The king sat on a rectangular saddle cloth that had tassels along its lower edges while mounted hunt servants and cavalrymen sat on folded over animal skins.

The earlier crests of the reign of Ashurnaṣirpal II are different from the later ones of Ashurbanipal but they are not described here as they did not affect the way the horses were handled. A caparison that would give a horse some protection against arrows or animals' horns or claws was worn by a hunt servant of Ashurnaṣirpal but did not become general in Assyria's cavalry equipment issue for cavalrymen till Ashurbanipal's time.

Ashurbanipal's relief illustrating his great victory over the Elamites at the Battle of the Ulai River in 655 BC gives us some indication of Assyria's enemy's equestrian equipment. It shows the battle, of course, as a complete victory over an inferior and less well equipped foe. Of the Elamite chariots, only the wheels with their twelve spokes remain in the river, the chariot bodies having been destroyed in some way. The Elamite horsemen have bridles and reins similar to those of the Assyrians but they are shown as riding bare backed, which cannot reasonably have made them much inferior to the Assyrians as horsemen taking into account what a threat they had been to the Assyrians for many a year.

It is not immediately obvious why the Assyrians should have increased the team of horses pulling a chariot from two to four in the reign of Ashurbanipal. It could have been because the real threat of a chariot was the width of the team of animals drawing it and by increasing the team from two to four widened the path that the chariot would carve through an enemy force by another two metres. But this is supposition. For ancient kings or commanders of chariot forces have never said that this rather than the alarming appearance of a chariot bearing down on the enemy lines or the missile power of its crew was the effective weapon of this vehicle. But increasing the number of horses did not add to the number of arrows being loosed from it, it merely increased the risk of shooting your own horses. In any case no additional archers were added to Assyrian chariot crews.

It could be because the extra horses increased the range of a chariot. The extra horses could take it further into enemy territory before succumbing to fatigue. But in the late 7th century BC the Assyrians were not operating far away from their metropolitan base. There was no significant change in the type of horses pulling the chariots except for a possible increase in the height of chariot horses by one hand, four inches, due to better feeding and selective breeding, as suggested to the author by Mary Aitken Littauer in a personal communication in the 1960s.

No. What is observable is that in the late days of the Assyrian Empire the Assyrians were operating against enemies such as the Elamites, the Urartians and the Scythians who were uncomfortably close to the borders of Assyria and lived in rocky mountainous terrain. What the two extra horses would have done was to make it easier to pull the chariots uphill into enemy territory, assuming that at this juncture the Assyrians were using chariots at all rather than cavalry and keeping the wheeled vehicles for show. I suggest that the Egyptians and armies of other countries in Palestine and Syria did not increase their chariot teams from two to four because in the more level plains in which they operated they did not need to.

The design and choice of ancient bits and bridles has been the subject of much experimentation throughout the centuries. As already stated the commonest bit used in antiquity and indeed today in the West is the jointed snaffle that hinges in the middle. The way in which the Assyrian bits were attached to the bridle differs slightly from modern Western ways. In modern Western tack the rings on the ends of the bit hang from straps called 'cheek pieces', that themselves are attached to the strap that go over the top of the horse's head. They are not attached to the nose band. With Assyrian bits, metal cheek pieces on the ends of the bit are suspended from Y-shaped divided ends of the cheek pieces and are stitched to the nose band. The 'cheek pieces' rise up to meet the strap that comes up the other side of the horse's head. The Assyrian nose band does not go round the horse's nose, but each side of it is slanted up towards the middle of the animal's face and they would meet high up on the horse's face where there must surely have been another strap coming down the centre of the horse's forehead to hold this nose band in place. I call this a nose band because it takes the place of the nose band on modern European tack, but it does not actually impinge on the horse's nose. So it cannot have had much of a restraining effect on the horse's head.

It is not at all easy to say how the reins were attached to the bits. This is probably a point where the highly detailed portrayal on the palace reliefs of the Assyrian equipment was at fault. Often the reins disappear behind the metal cheek pieces of the bits, which is unexpected, and would suggest that the bit did not protrude out through a hole in the metal cheek pieces as with the early 'run-out' bits in Siberia. The bit on a horse on a relief of Ashurbanipal (relief number BM 124 86 of room C) is more precise and shows the end of the rein threaded through a hole in a short metal or leather strap that goes into the horse's mouth behind the curved metal cheek pieces of the bit. This suggests strongly that the bit does not go through these cheek pieces as one would expect.

Exceptions to the general rule in the Middle East, leaving aside Assyria itself, that two horses were sufficient to pull a chariot were Mycenaean Greece and China where occasionally two more horses were added to the team. This augmented Assyrian team probably did not become common until the time of Sargon II (721–705 BC) or Sennacherib (704–681 BC), the third last important king of Assyria. Ashurnaṣirpal II, the first great Assyrian monarch, started his reign by attacking the hill men in the on the foothills of Anatolia in the area north of the River Tigris and the Euphrates.

On ascending the throne, Sargon's first preoccupation was with suppressing a rebellion among the Assyrian provinces and satellites in Syria. Then in his eighth year he mounted a powerful attack on Urartu, a rising state that had chariots and was becoming an important power in the region round Lake Van in Anatolia. This was because of its control of the valuable east-west trade route that went along the passes between the mountains. We know more about Sargon's eighth campaign than we do of most other Assyrian annual military operations. For instead of the details of the operation being confined to the usual mention in the king's annals, it was the subject in this year of a special letter that Sargon wrote to the national god Ashur in which he gave more than the usual details of what had happened. Here he particularly wanted to assure the god that he had fulfilled his command and that he led an army on the usual annual campaign against enemies of the god's country, Assyria.

The country round Lake Van is extremely mountainous, with the valleys running east and west, at right angles to the route that an invading Assyrian army would have had to take. Mentions in a cuneiform text that the chariots had to be hauled up the slopes with ropes show that it is not chariot country. Even with an augmented team of four horses the Assyrians must have had extreme difficulties in hauling their vehicles up the inclines, although of course overcoming these difficulties was presented to the god as great achievements, which in truth they were.

Chapter Eight

Elam

The kingdom of Elam consisted of two regions, the plain of Susa, south-east of Mesopotamia, and the highlands east of Assyria. The people, the Elamites, spoke a language that has not yet been translated and although it was written in a form of cuneiform it has no connections to Sumerian, Semitic languages, or the Indo-European languages of the Iranian highlands.

The Elamites first appear in the historical record between about 3200 and 2700 BC in as yet untranslated texts in Proto-Elamite script from the important city of Susa that preceded their adoption of cuneiform. This was followed by the Old Elamite and Middle Elamite periods of the late 3rd millennium to the late 2nd millennium BC.

From the Middle Elamite period onwards the Elamite urges to expand into fertile Mesopotamia brought them into conflict with first the Sumerians and then the Semitic Akkadians who occupied central and southern Mesopotamia after the demise of Sumerian power and before the arrival of the Assyrians and Babylonians.

The Kassite dynasties who ruled Babylonia in the early 2nd millennium BC controlled Elam until about 1210 BC when Kassite power collapsed and Elam attained the zenith of its power. Elam succeeded in conquering Babylonia. Then Elamite power suffered a decline for the three hundred years until about 770 BC.

At that time Indo-European speaking Iranian people were migrating from the Iranian plateau into the eastern border lands of Assyria that the Elamites occupied. Relations between Elam and Assyria were complicated but never cordial and their disagreements often resulted in Assyrian military campaigns to keep the Elamites out of the Assyrian plain. This was the time when Sennacherib and Ashurbanipal were experimenting with teams of four horses that could haul chariots up the steep and rough hills that the Assyrians had to cross to campaign in Elam.

In 653 BC Ashurbanipal won a great victory over the Elamites at the Ulai River, the most detailed of all the Assyrian battles. It was portrayed in the reliefs in the palace of Sennacherib in Nineveh. In a most dramatic way the Assyrian forces, whose leading units at the point of contact were mounted horse archers and spear men, cavalry rather than chariots, forced the Elamites, chariots and riders, into the water. All the characters in the relief, both Assyrians and Elamites, are shown moving to the right so that only their right hand sides are visible. The Assyrians do not seem to have ridden their mounts into the water, but dismounted and pursued the retreating mounted Elamites into the river on foot. The Elamites are shown as so disorganized, with their men riding for their lives, that it is not possible to say whether the Elamites were disciplined cavalry or just individual riders. But the Assyrians were definitely organized disciplined cavalry.

It is clear to a rider studying the large relief that both of the Assyrian mounted archers' hands are visible, holding the bow and arrow. This is not the case with the Assyrian spear men. For while the right hand of the mounted spear man is seen to be grasping his spear over hand, his left hand cannot be seen as it is holding the reins behind the left side of his horse's neck. The significance of this when charging at an opponent is dealt with in detail in chapter 7. Here, let us just say that the relief in this instance portrays accurately the problem involved in getting a horse to charge at someone and demonstrates that even although they had neither saddles nor stirrups, the Assyrians were capable horsemen with a thorough understanding of a horse's balance and how the rider should manage it to advantage.

The Elamite chariots are noteworthy in having twelve or sixteen spokes on their heavy wheels and in having no protective walls, openwork or solid, round their fronts and sides. Their chariots consist of just flat openwork platforms, perhaps with woven mats on the floors. The only construction they have above the floors are single semi-circular hoops on the sides that would have protected the seated driver and the three archers of the crew from getting an arm or leg caught between the rotating spokes of a wheel. The axle is under the centre of the body.

The Elamite horses, two or four in number, have head collars but no bits or reins. They are guided by a long whip held by the driver who gives them taps on the head. This cannot have been very effective in a crush of wheeled vehicles.

This chariot that belonged to the Elamites is so unusual, so technologically inefficient and archaic, and unlike any other horse-drawn fighting vehicle, that one must feel compelled to ask what are its antecedents. Its unusual constructional feature is the large number of spokes in its wheels. Whether they used light chariots like the Egyptian vehicles or heavy ones like the Assyrian version, other cultures in the Near East found it desirable to have the chariot body surrounded by a guard rail or wall that was at least thigh high, against which the crew could brace themselves and behind which they could stand to drive the vehicle and fight from it. But we see clearly from the Assyrian relief that the Elamite drivers drove sitting down. It is reasonable to assume that the warriors in the crew used their weapons while seated lest they be pitched overboard and relief number 124939 in the British Museum supports this suggestion. There was nothing like a guard rail or wall to save the men aboard, only the low hoops to save them from being caught up in the revolving spokes.

The only other chariots that show these features are those excavated at Lchashen in Armenia and Anyang in China, and the Chinese ones are quite definitely derived from the Armenian ones. They are dealt with in detail in chapters 2 and 13. But in 1200 BC the Armenian and Chinese ones are much earlier, probably by around six hundred years, than the ones the Elamites had at the Ulai River. In spite of this it is conceivable that the Elamites were using chariots of a very archaic Central Asiatic design of which no representations from that region survive. If this suggestion of mine gains acceptance it will be a significant advance in the evidence for very early chariot design in Central Asia from before the adoption of the chariot into the military inventories of Near Eastern countries.

Chapter Nine

Egypt

Egyptian political history is divided by Egyptologists into kingdoms and dynasties. The Old Kingdom, which is composed of the Ist to the Vth Dynasties, lasted from 2686 to 2181 BC. Then came the First Intermediate Period from 2081 to 2055 BC and after that the Middle Kingdom of the XIth to the XIVth Dynasties, which lasted from 2055 to 1650 BC.

During the Old and Middle Kingdoms the Egyptians followed a policy of exclusion from foreign contacts in Asia. For in Egyptian eyes they had constructed the ideal society, so why risk change by exposing it to foreign influences?

In the Second Intermediate Period between 1650 and 1550 BC Egypt was taken over and ruled by a grouping of foreign immigrants generally known as the Hyksos. They entered Egypt from the Levant around 1800 BC and gradually gained political power until they ruled the country. They were a coalition of peoples of Palestinian origin, although whether the language they spoke belonged to the Semitic or Hurrian family is disputed.

The story of Egypt's wars with its large chariot army starts about 1674 BC, at the end of the Second Intermediate Period in Egyptian history, between the Middle and the New Kingdoms. The Hyksos had been infiltrating Egypt since the Xth Dynasty in the 22nd century BC. This was during the period of Egyptian weakness in the Middle Kingdom. Something that is obvious to modern Egyptians, and whose memory is immensely irking to them but rarely appreciated by outsiders, is for how much of their history Egypt has been ruled by foreigners, Hyksos, Sea Peoples, Greeks, Romans, Turks, Albanians and British before an Egyptian, Colonel Nasser, came along in the 1950s.

The Hyksos introduced many new things into Egypt such as hump-backed zebu cattle and new vegetable and fruit crops; the compound bow, which had a much greater range than the earlier bow made of a single piece of wood; superior bronze casting; and also the chariot. This was probably a Canaanite invention, which they brought with them from their former abode on the eastern Mediterranean coast.

The Hyksos were expelled around 1560 BC by the first Egyptian pharaoh of the XVIIIth Dynasty, Amosis. This was now the New Kingdom, which included the XVIIIth, XIXth, and XXth Dynasties and lasted from around 1570 to 1085 BC. In the XVIIIth Dynasty, following the Hyksos' rule, the Egyptians, with an eye on the wealth passing along the trade routes through Palestine, realized that the Old and Middle Kingdom beliefs that Egypt's position was cushioned by deserts from the dangers of foreign invasions were no longer tenable. Also, because of the increasing power of the Hurrian Kingdom of Mitanni in Syria, the Egyptians decided that they could no longer afford to stay out of affairs in Asia. So between 1525 and 1512 BC the third pharaoh of the XVIIIth Dynasty, Tuthmosis I, led his army into Palestine to confront Mitanni.

The 2nd millennium BC saw a series of bitter wars in the eastern Mediterranean seaboard for control and domination of valuable trade routes both from the north and particularly from the east into the Levantine coast and its hinterland. They threatened the interests of Egypt as well as those of the Neo-Hittite survivors in north Syria of the now collapsed Hittite Empire in Anatolia. They also involved the states in central Palestine and Syria of other peoples such as the Philistines, the Hebrews, the Canaanites, and other inhabitants of Syria such as the Amorites and Aramaeans who found themselves caught in the middle of wars not of their own choosing. That is not to say that these lesser states caught in the middle of international power struggles were innocent victims. They could be equally expansionist when it suited them. Meanwhile, the Assyrians were a growing power and menace in the east. The difference between defence and aggression is a very fine one.

In a resurgence of Egyptian strength after the weak Middle Kingdom it was then realized by her rulers that if Egypt were to remain prosperous and independent she could no longer afford to stay shut off from the rest of the world in the 'perfect society'. She had to control the vital trade routes that crossed the Levantine seaboard. So for the next four hundred years Egypt was ruled by Egyptians and pursued a forward foreign policy.

Left: Illustration of Sumerian pictogram of sledges and sledges with wheels or rollers from Uruk level IV a. The upper two rows are of sledges. The lower row shows sledges with wheels or rollers under them. (*Author's drawing*)

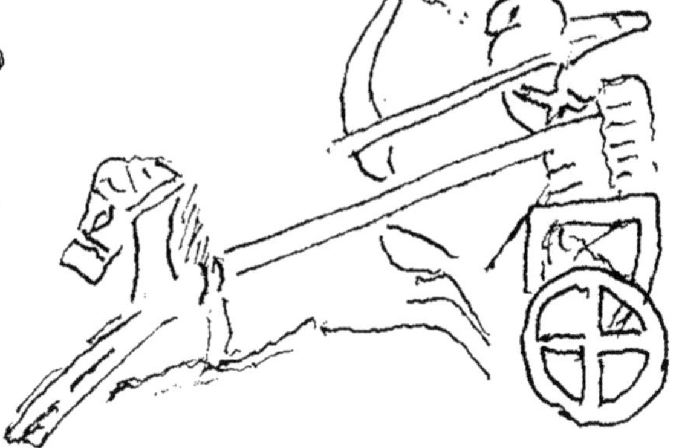

Right: The chariot on the golden bowl from Ugarit. Louvre AO 17208. No photograph of the bowl shows the chariot clearly from one viewpoint so it is shown here in a drawing. This is an image of possibly the earliest type of Middle Eastern chariot. (*Author's drawing*)

The Standard of Ur. The War Panel showing the battle wagons charging. (© *Trustees of the British Museum*)

Close-up view of two battle wagons on the Standard of Ur. (© *Trustees of the British Museum*)

The Wheels of War project. The author driving a battle wagon at speed with four donkeys pulling it. (*The Estate of the late A. Cernock*)

The Wheels of War project. The battle wagon going at speed into a turn to the right. The excitable donkey is on the far left. (*The Estate of the late A. Cernock*)

A cuneiform tablet of the Birth Legend of King Sargon II of Assyria. (© *Trustees of the British Museum*)

Modern gilt replica of a chariot from the tomb of the Pharaoh, Tutankhamun. This shows how light the wheels were. (*Fotolia*)

Rameses III charging in his chariot. This shows how small, light and top heavy the vehicle was. This illustration is probably reasonably accurate, even though it is an official wall painting that glorifies the pharaoh. Bronze Age armies were led by the commander from the front. (*Fotolia*)

Early Assyrian chariot from a seal impression of the reign of Ninurta-Tukulti-Ashur (c. 1133–32 BC) of Assyria. (*Author's drawing after Madhloom*)

Assyrian chariot of Ashurnaṣirpal II. Ashurnaṣirpal hunts lions from his chariot. Throne Room of Nimrud North-West Palace. (© *Trustees of the British Museum*)

Tiglath-pileser III of Assyria in a procession in his chariot. (© *Trustees of the British Museum*)

Assyrian cavalrymen using lances in battle. (© *Trustees of the British Museum*)

The Assyrian King Ashurbanipal is shown shooting an arrow from horseback during a hunt. (© *Trustees of the British Museum*)

Relief from Carchemish showing a Neo-Hittite chariot. (*Courtesy of the Museum of Anatolian Civilizations, Ankara*)

Battle of Til-Tuba (The Ulai River). This shows the peculiar Elamite chariot that has no side walls. (© *Trustees of the British Museum*)

Assyrian riding horses harnessed up for King Ashurnasirpal to go hunting. (© *Trustees of the British Museum*)

Assyrian horse archers in battle. (© *Trustees of the British Museum*)

An essential for this was an efficient army and in the 2nd millennium BC the basis of that was a highly mobile arm of chariotry. The Hyksos had made that possible for the Egyptians.

The art of riding on the backs of horses was not yet developed enough for horses to be used as massed disciplined cavalry in warfare. So for military purposes the Egyptians were forced to confine riding horses to the few messengers necessary to provide communication in a Bronze Age army.

We have already stated our belief that the chariot was a Canaanite invention, that the Egyptians learned of it from the Hyksos. In support of this is the fact that the Egyptian name for a chariot in their Kushitic language, *merkebet*, was of Semitic Canaanite origin. It is similar to the word for a chariot in other Semitic languages. No military innovation, be it chariots or tanks, takes longer than it takes a man to return home from the front with the intelligence that the enemy has a new surprise weapon that could be adopted to its advantage by his own country.

The evidence for chariots as distinct from agricultural carts and wagons and the lumbering Sumerian battle wagons appears in the Near East first in Canaan. There is a golden bowl of about 1400 BC excavated in the site of the ancient city of Ugarit in Syria and now in the Louvre that has an embossed hunting scene with a chariot in it. This is a very light box-like horse-drawn vehicle with two four-spoked wheels whose axle is under the centre of the body. It carries one man who has not much more space than the part of the floor on which he stands. He is shooting arrows while he has the horses' reins tied round his waist, a most dangerous procedure but understandable under the circumstances. The horses being only small ponies of thirteen hands in height would be an advantage, for the danger in using any weapon, be it bow and arrow, sword, or polo stick on a horse is that the horse might be hit. The smaller the animal the better.

The sides of the Ugarit chariot body are panelled, as were the early Egyptian ones. There is a wall painting from the tomb of Userhet, royal scribe in the reign of Pharaoh Amenhotep II of the XVIIIth Dynasty and dated to about 1420 BC, which has a hunting chariot scene on it, closely similar to the one on the Ugarit bowl. Egyptian hunting chariots of this early part of the 2nd millennium had a case for the bow and an arrow quiver attached to the panelled sides of the vehicles. It did not take long for the

Egyptians to move the axle from the centre to the rear of the chariot body and increase the number of spokes on the wheels to six. After an experiment under Amenhotep II (1436–1411 BC) with a chariot wheel of eight spokes the Egyptians settled in the reign of Thutmosis IV (1411–1397 BC) for a wheel with six spokes.

From surviving chariots in tombs we know that under Tutankhamun (1361–1352 BC) the body of a chariot could be 1.25 metres high, 1.02 metres wide and the central draught pole be 2.56 metres long. The track of the 0.92 of a metre diameter wheels was 1.75 metres. The six-spoked wheels of a chariot found in a tomb in the Valley of the Kings, and now in the Cairo Museum, belonged originally to Yuya, father-in law of Amenhotep III, also known as Amenophis III, (1417–1379 BC) and his wife Thuyu. These are smaller at 0.74 of a metre but that does not change our general idea of what Egyptian chariots were like.

Egyptian chariots had yoke braces, a feature not found on other chariots. These were leather thongs branching off midway along the draught pole and attached to the yoke to keep it at a right angle to the pole. Yoke saddles, a feature of Egyptian chariots but not of Western Asiatic ones were an effort to adapt the yoke, originally intended for bovine traction, to something more suitable for equids that did not have a hump.

Surviving bits from the Near East as well as Egypt and Transcaucasia were of bronze, although antler or bone cheek pieces that must originally have had rope or leather mouth pieces occur in Anatolia. They date from after the middle of the 2nd millennium.

When the chariot reins are tied round the pharaoh's waist in wall paintings careful study shows that the right reins are lashed to the right side of his hips and the left reins to the left. So presumably he could exercise some steering by turning his hips. But the practice of lashing the reins to the driver's body is not one that would recommend itself to modern horse drivers.

Egyptian chariots had a two-man crew, driver and archer. In records of Kadesh a shield is held by the archer when advancing into battle, but in action is handed to the charioteer. He seems to have reached forward to pluck at the reins which are tied round the archer's waist.

From the time of Seti I Egyptian chariots carried a pair of javelins as auxiliary arms.

We are fortunate that the climate in Egypt is so dry that actual chariots have survived intact and in good condition. For a glimpse of what chariots from other Middle Eastern countries of the middle of the 2nd millennium BC looked like, we are dependent on Egyptian wall paintings or reliefs carved on stone. Even with the heavy Assyrian chariot, whose side-on view we know so well from the numerous likenesses in their detailed reliefs, not a single actual chariot survives. And for the appearance of chariots belonging to the Assyrians' enemies we have only occasional glimpses of them when the vainglorious Assyrian kings thought it would add to their prestige to display their enemies in defeat.

The frames of the Egyptian chariots found in tombs were of ash or elm, bent with heat or steam into shapes that had few joints and were therefore elastic and resistant to collapse under strain. The fronts of the bodies were covered in ox-hide and the sides were left open. The draught poles were of elm and had at their fronts a yoke of willow. The floors were D-shaped, being curved at the front, and the wheels were constructed in a peculiarly Egyptian manner that combined extreme lightness with strength. The usual way of constructing a wooden vehicle wheel, one that survived to the 20th century CE, when wooden wheels gave way to metal ones, was for the spokes to be inserted at one end into sockets in a wooden hub and at the other end into sockets in the several felloes that together made up the circumference of the wheel. The felloes were held together and protected from abrasion by a tyre, often of metal. The tyre was constructed in the earlier years of this current era of several pieces and later on of one ring that was shrunk on round the felloes.

The Egyptian chariot wheels were made in a way that dispensed with the heavy felloes. With little protection for the wheel rims they must have needed many running repairs and have been difficult to work on, since repairs would have necessitated complete dismantling of the wheel. The felloes and hubs were cut from ash and the spokes from almond or plum tree wood. The spokes were made from pieces of wood selected for having a natural V-shape so that each limb of them formed half of two separate spokes. These were glued together and the sharp turn of the V-shape was attached to the hub with wet cattle intestines that shrank and hardened as they dried. The spokes fitted into sockets in the light felloes and there were outer felloes that were the running surface. The whole rim was lashed to the spokes with bronze wire and a raw hide tyre was shrunk on to hold the wheel in shape. This wheel had considerable resistance to bumps but how well it survived the abrasion of lengthy journeys we do not know.

The Egyptians' Canaanite enemy was still driving chariots with four spokes to the wheels and the axle under the centre of the body. But war accelerates the speed of the adoption of new technology and it did not take the Canaanites very long to adopt the new Egyptian type of wheel, although they seem to have kept the old position for the axle under the centre of the body.

Of course we are dependent for our knowledge of the development of the Canaanites' chariots on the Egyptians' portrayal of them, which only occurred at long intervals. It would be surprising if the Canaanites did not upgrade their chariots much more often than the pictorial evidence informs us. The Canaanite retention of the central axle also brings into question the extent to which under battle conditions with a very light vehicle the position of the axle really affected the turning circle. We do not know, but experience with a replica Sumerian battle wagon suggests that if the animals are galloping they can drag the vehicle round in a turn so tightly that the resistance of the wheels and axle to being dragged sideways is more important to the chariot's stability than the actual position of the axle under the chariot body.

An argument as to what is the tactical use of the chariot, to charge straight at the enemy and crush it or to serve as a moving platform from which missiles in the shape of javelins or arrows can be launched, must depend on whether the principal weapon is the team of horses charging onto the enemy or the missiles that can be launched while the chariot is moving at about twenty miles an hour across the face of the enemy army.

One thing is certain. A frontal chariot charge can only succeed when there are gaps in the enemy forces through which a chariot can be forced, or the enemy is so demoralized by earlier attacks that its men will turn and run as the chariots approach. The same applies to cavalry charges. The psychology of the horse, an animal whose only defence is flight, has not changed over the centuries.

The smallest unit of Egyptian chariots known to us was five and we are justified in assuming that Egyptian chariots fought in such units or as part of the next larger unit of ten chariots under the overall command of a *kedjen-tep*, a first charioteer. Above that troop of ten chariots was a larger squadron of fifty chariots. Performing their coordinated manoeuvres at speed must have taken a great deal of training and we know that chariot warriors were exercised in engaging targets with a spear at speed. Unfortunately, lack of resources has restricted all modern re-enactment with chariots to experiments with a single vehicle. In real combat a single chariot facing the massed missile power of any ancient army would have been on a suicide mission. Accompanying each Egyptian chariot was a runner, a foot soldier who ran alongside it and provided local protection, as a destroyer protected a 19th century battleship.

Egyptian chariots were light with only a driver and a warrior as crew and therefore fast, although the best speed of a chariot depends on the body weight and length of legs of the horses, not principally on the construction and weight of the vehicle. There was no protection from the chariot for the crew above thigh height, when the warrior was dependent on the armour he wore. That armour worn could be of linen stiffened with resin or lamellar plates of bronze or leather. There existed light and heavy Egyptian chariots but the difference was in the weight of the protective armour worn by the crew, not in the weight or size of the vehicles.

It is reasonable to presume that early on in the battle, before the enemy was in flight, the task of the chariotry was skirmishing, to weaken the enemy's resolve and prepare for the final onslaught with chariots accompanied by infantry.

Following the principles of mobile warfare with horses that have stood the test of time, the individual troops of the Egyptian chariot arm would have approached the field of battle in line ahead, behind each other. As there would have been hundreds of chariots in the army the massed chariots would possibly have looked to the observer more like a square than a line, with many columns of troops each following their leader. Ideally they would have changed front into a line of battle with a long reasonably thin line of chariots facing the enemy before discharging their missiles and turning away. They would not have charged through the enemy line unless it was already severely shaken.

We do not know exactly what drill movements the Egyptians used or what formations they adopted in making their skirmishing attack. It could have involved each chariot taking its own course and attacking a target of opportunity at will. But a large number of vehicles moving at speed on courses that had not been pre-arranged must surely have led to an unacceptable number of collisions. It would have been much safer if the chariots had approached the enemy at speed behind each other in line ahead, before turning away to left or right in the classic cavalry volte once the enemy was within range so that the warrior could start shooting with his bow. The direction in which they would turn would be the one which in Egyptian experience gave the archer the clearest view. Then they would circle back to their own lines. Their only real protection would be their speed, for once the enemy was within range of them, so were they of the enemy, and he would for the present outnumber any individual chariot.

Fortunately for those who plan wars, the combats are fought by young men who each believe that it is the enemy who will be killed and not them.

The best course we can follow in our efforts to find out how the Egyptians used their chariots in battle is to examine the records of them. The best documented Egyptian battles are those of Megiddo (c. 1460 BC) under Tuthmosis III and Kadesh (in 1275 BC) under Ramesses II.

The Battle of Megiddo was the culmination of the most famous event of the campaign of Tuthmosis III of the XVIIIth Dynasty into the Eastern Mediterranean seaboard. He had decided that Egypt must exert its power in an area through which the east-west trade of the Orient passed. Megiddo was also, with the later battle of Kadesh, one of the only two Bronze Age battles in which the tactical use of chariots was described. Tuthmosis was opposed by a coalition of Canaanite states from the north of that region on the River Orontes. These were led by the King of Kadesh, who was supported by the Mitannians. At a daily rate of fifteen miles, which slowed to eight as he reached the territory of the enemy, the Canaanite princes, Tuthmosis led his army of fifteen to twenty thousand men and probably a thousand chariots through Sinai. They reached Gaza in the land of the Philistines in nine days: a difficult march through arid country.

The crucial settlement in the path of the Egyptians was the town of Megiddo (Armageddon in Hebrew), which was strategically situated in the Plain of Esdraelon, west of the range of the Carmel Mountains that runs north and south parallel with the coast and which shuts off the coastal cities from the hinterland. Megiddo's strategical significance was that it lay opposite the mouth of the pass of Aruna (modern Wadi 'Ara) which was a way from east to west through the difficult mountains. This pass was part of the main road that connected the coastal plain of Palestine, and therefore Egypt, with Mesopotamia, Syria and Anatolia. There were three routes by which an army could approach Megiddo, near which a large Syrian army was waiting for the Egyptians.

The Canaanite strategy was offensive, to seek out the enemy before he could be seen, while their tactics, once contact had been made, were defensive. So at the approach of the Egyptians they retired within Megiddo, which the Egyptians besieged for some months before finally taking it. Canaanite tactical dogma was to use heavy chariots and crews wearing armour, particularly the chariot-owning Mitannian aristocracy, the *maryannu*, to skirmish ahead of the main infantry army and wear down the enemy with missile attacks with arrows. As far as we can judge the heavier Mitannian chariots were not as fast as the much lighter Egyptian ones, but they could withstand battle damage better. Surprise attacks and outflanking movements could be undertaken when the opportunity arose. A chariot attack on an approaching enemy army who is stretched out in a long column while approaching the site of battle was a good move.

The Canaanite army was positioned to cover the north and south points of entry into the plain of Esdraelon. Instead, at least in the official account, Tuthmosis personally overruled his generals who favoured an open approach across the plain. They considered that the Pass of Aruna approach was too dangerous because it was so narrow in places, no more than 30 feet wide at the most, that the chariots would have to drive through it in single file and be exposed to possible ambush. Of course the art of 'spin' is not new and Egyptian, and other, royal accounts all emphasize the personal valour and intelligence of the ruler.

On the strong pleas of his generals, Tuthmosis agreed to wait where he was at the mouth of the Pass till the tail end of his army caught up with him. That has the ring of truth. It took the whole Egyptian army twelve hours to make its way through the Pass of Aruna and it was early evening of that day before the last units of it emerged from it onto the plain. There they found themselves a little north of the Canaanites who, taken by surprise, rushed back into Megiddo.

The following morning the Egyptian army was drawn up before Megiddo, ready for battle. It advanced. The remaining Canaanite forces broke and ran for the shelter of Megiddo, a retreat that turned into a rout as equipment was abandoned. The equally panic stricken inhabitants of Megiddo closed the city gates on the retreating Canaanite soldiery. Canaanite kings shut outside had to suffer the indignity of being hauled up to safety inside the city by sheets let down from the walls. It should have been an easy and complete Egyptian victory but that was prevented by the Egyptian soldiers pausing in their advance to loot the equipment left behind by the retreating Canaanite army. So Megiddo had to be besieged. The Egyptians set about that in a thoroughly professional manner. A moat was dug around the city and beyond that a wooden palisade was erected to prevent a break out. The siege lasted seven months, after which Megiddo surrendered. The Egyptians took vast booty, pride of place going to 2041 horses, which were taken back to Egypt to be used as breeding stock.

The remaining eighteen years of Tuthmosis' reign saw him campaigning almost every year in Asia in a constant endeavour to preserve Egypt's trading position in the Levant in the face of the expansionist mercantile ambitions of Assyria in the east and the Hittites in the north. He had managed to reduce the influence of Mitanni to a level with which Egypt could cope and there followed a period for the next hundred and fifty years of what passed for peace in Syria, meaning that in spite of continual small wars of limited success between second rank states, there were no large trade wars or battles between major powers. They were all involved in their own internal affairs for a considerable part of that time. In those days of absolute monarchs the international rating of a country depended on the personality and acquisitive aggressiveness of the ruler and whether he was intent on advancing his country's position in the region or whether he was distracted from the pursuit of international prestige and trading success by pressing internal problems.

The major powers, Egypt, the Hittites and the Assyrians, were still jockeying for pre-eminence but none of them had reached a position where it was able to alter the regional balance of power. Assyria was still in effect a client of

Mitanni while Egypt was suffering from the weakening isolationist policies of the pharaoh Amenophis IV (1379–1362), better known as Akhenaten. His efforts to introduce the newly invented religion of the worship of the Aten, the sun disc, to replace the complicated and inconsistent existing Egyptian religion, upset the whole religious, philosophical and administrative base of the deeply conservative Egyptian state in which the priesthood had a position of power to defend. So Egypt was not in a condition to come to the support of its Syrian vassals when they were threatened.

The other major power, Mitanni, was in the throes of a civil war. The current king, Tushratta, had come to power by murdering his brother. His family was not pleased with that and another brother declared himself a rival king. In the usual palace intrigues the anti-Tushratta party sought the support of the Assyrians and in 1350 BC Tushratta was murdered by one of his sons. The legitimate heir to the Mitannian throne, Mattiwaza, fled to Babylon where the king of that country, maintaining a policy of neutrality over other peoples' problems, refused him political asylum. So Mattiwaza sought refuge with the Hittites, who were not currently pursuing an expansionist policy. But right at that time, about 1389 BC, a new forceful Hittite king, Suppiluliumas I, had ascended the Hittite throne and this was bad news for Mitanni. The Assyrians, under their king Ashur-uballit I (1365–1330 BC) in concert with another unnamed small state that does not appear again in the history books, moved against Mitanni and in an apparently bloodless invasion took it over. That was the end of Mitanni and the beginning of Assyria as a major power. Ashur-uballit took the rather bombastic titles of 'Great King' and 'King of the Universe', addressed the pharaoh of Egypt in correspondence as his brother and gave his daughter in marriage to the king of Babylonia.

The political situation was, by the standards of the times, fast moving while the modern historian who is trying to disentangle the tortuous politics of the period has to take the occasional references to the situation in various texts and from them build up a picture of what was happening. Suppiluliumas entered a policy of military and political expansion that would place the unstable city states of Syria firmly under Hittite control.

The other important battle in which chariots were involved and of which we have a reasonably full description was that of Kadesh, also spelled Qadesh, in 1274 BC. It took place in northern Palestine between the Hittites and the Egyptians under the pharaoh Ramesses II in another of those trade wars that the Egyptians fought in an effort to control the valuable trade in luxuries from the east. The battle takes its name from the nearby town of Kadesh that lies on the Orontes river in Syria some little way inland and approximately half way between Damascus in the south and Aleppo in the north. The Egyptians were in the process of setting up camp to the north-west of the town when the Hittites and their local allies, who had been hidden from view beyond the east of Kadesh circled round the town, crossed the river, and delivered a smashing surprise chariot attack on the unprepared Egyptians, catching the Amun Division that was under the personal command of the pharaoh while it was still in line of march. Only the timely arrival from the south of two fresh Egyptian divisions and a Syrian allied force saved the Egyptians from a resounding defeat. Even although the Hittites had won this particular phase of the battle they were still called by the Egyptians in their later description of what they subsequently claimed as a great victory as 'The wretched foe'.

What had happened before the Hittite attack on the Egyptians was that two Arab bedouin, masquerading as deserters from the Hittite forces, had been brought to the pharaoh after being beaten and they told him a tale that the enemy was still in front of the Egyptians, far to the north, when in fact they were behind the pharaoh's forces. The pharaoh swallowed the tale and the subsequent Egyptian account of the battle emphasized the deceitfulness of the supposed deserters, not the pharaoh's lack of guile in believing them.

After their successful attack on the Egyptian column the Hittite chariot forces lost their advantage by stopping to plunder the Egyptian forces they had overthrown and they were in turn overwhelmed by the fresh Egyptian troops, who included a crack Canaanite unit in Egyptian service.

The main body of the Egyptians rallied, being rescued from their local defeat, and turned eastwards, towards the town of Kadesh, behind which were the Hittites with the River Orontes at their back. But the Hittite king had now committed and used up his chariot force and his 6000 infantry were no match for the fresh Egyptian chariots. The Hittites retired into Kadesh and prepared to be besieged. The Egyptian investment of Kadesh does not seem to have been successful and the battle ended in a stalemate after which the Hittites and the Egyptians signed a non-aggression pact, although the Egyptians unsurprisingly declared it a victory.

But the Egyptians did achieve political control over a large stretch of Syria that had previously been part of the Hittite sphere of influence and the battle of Kadesh showed that if properly handled chariots could be extremely successful in surprise attacks on infantry that was not prepared to receive them. It also showed, although whether this lesson was learned at the time is another question, that chariotry, like the later cavalry, was a one-shot weapon than could be used only once in any particular battle. We know from later wars that cavalry horses will charge at least once, but many will never do it again – wise animals.

The reliefs of Ramesses II (1304–1237 BC) of the Battle of Kadesh show the Hittite chariot warriors with lances but in the reliefs of Ramesses' father, Seti I (1317–1304 BC), they are archers. The use of chariots to carry archers seems to have been common throughout the Near East in the Late Bronze Age. Exceptions to this practice were the Hittites and the Mycenaeans, both Indo-European language speakers, who armed their chariot warriors with spears.

Egyptian, Ugaritic and Hittite records mention the hiring as mercenaries of Hapiru, (*'prw* in Egyptian), a name that comes from a Semitic verb meaning to wander. The similarity of this name to the much later name of Arabs is obvious. They must have been Semitic nomads, ancestors of the Arab bedouin. They seem to have been hired as common foot soldiers and not as skilled chariot crews.

Another enemy of the Egyptians who merit a mention here but are not important enough to be given a section of this chapter on their own are the Libyans who lived in the desert to the west of Egypt and had been a thorn in the Egyptians' side for a long time. In 1185 BC Ramesses III killed or captured 4200 of the Meshwesh tribe of Libyans along with 93 chariots and 183 horses. So by this date these Libyans had imported chariots and horses and had gained the skills to use them.

Chapter Ten

Palestine, Syria, and Cyprus

The political and military history of Palestine's, Syria's, and Egypt's involvement in wars in the Near East in the 2nd millennium BC is an account of a jumble of medium sized states who were perpetually forming alliances, breaking them, and going to war with each other.

Many of the comparatively small countries there had sizeable chariot arms in their forces, but unless they went to war with Egypt and their forces ended up being illustrated on Egyptian reliefs or wall paintings we do not know what their fighting vehicles looked like. Therefore we are unable to assess their effectiveness and determine how they were used in action.

Hebrews aka the Israelites

With terms such as Hebrews, Israelites, Jews and Judeans, which do not exhaust the whole catalogue of names used for the people who escaped from bondage in Egypt, we have enough names for the historian to wrestle with.

Fortunately, historians who write in the English language have come to a general decision. Hebrews are what the ancient Egyptians and a few other peoples who disliked them called the group of many different tribes of Semitic language speaking people who left Egypt for Sinai in about the reign of the pharaoh Ramesses II (1304–1237 BC) or shortly afterwards. So that is what we will call them here. They then appear in the Old Testament under the name of Israelites in the Second Book of Samuel, written long after the event, and once they have left Egypt and are wandering in the Sinai Desert we call them the Israelites. The exact relationship between Hebrews and Israelites, whether they are the same people, or a slightly different selection of the tribes who undertook the Exodus from Egypt together, is not at all clear and is a subject best left for deep academic discussion. But the term Hebrews seems to have disappeared from use by the time that they had arrived in Canaan and established a monarchy with Saul, the first king of the United Kingdom of Israel in the 11th century.

The biblical forty years that the Children of Israel spent in the Wilderness, during which time a collection of wandering nomadic tribes was welded together into one people with a single monotheistic religion, would actually have been much longer, more like several centuries. It is best to remember that at this early stage the followers of Moses were a great conglomeration of different tribes who were slowly becoming a people.

By the time Joshua led the various tribes into Canaan in the Late Bronze Age before the establishment of the monarchy they called themselves the Israelites. The term Israeli is a post-1947 CE one, referring to the citizens of the modern State of Israel and is not used in studying or writing about ancient history.

Canaan was the southern part of what later became known as Palestine and happened to be a highly sensitive region where the political and economic interests of three major regional powers, the Egyptians, the Hittites and the Assyrians converged.

The ten semi-nomadic tribes of the Israelites adopted a new religion with a single god, Yahweh, whose name was later transliterated, particularly in Western Europe, as Jehovah. This monotheistic religion was moulded during their wanderings in the Wilderness into Judaism, the Jewish religion. They had a covenant with him. They would worship only him and in return he would provide them with a country in Palestine, which would be theirs alone. Jews are followers of the Jewish religion. There are Jews today who are atheists and do not practise the Jewish religion, but still regard themselves as Jews. Here they are making an ethnic distinction, not a religious one.

With the ingrained sense of superiority over the effete townies that all nomadic peoples feel, the Israelites experienced a big shock when they discovered that the urbanized Canaanites whose land God had promised to the Children of Israel were culturally more advanced than they were. But they coped with this and occupied Canaanite towns, while substantially avoiding being contaminated by the former inhabitants' religion. For the first century after 1200 BC the tribes of Israel and Judah were scarcely urbanized and had no centralized government before the monarchies of Saul and David.

On arrival in Canaan the Israelites had a loose form of government under political/religious leaders called Judges. But they were surrounded by other Semitic language speaking peoples, plus the Egyptians and the Philistines in the south, who were as equally expansionist as they were. So warfare was continuous. The fighting men of the Israelites came from tribal levies, whose availability in time of danger was dependent on the shifting currents of tribal politics. A more permanently available army under a fixed and accepted supreme commander was needed.

So despite warnings from the Judges that the Israelites would live to regret having accepted the authoritarian rule of monarchs, in the 11th century the Israelites established a monarchy with Saul as the first king of the land they occupied west of the Dead Sea and the River Jordan. This became the United Kingdom of Israel. The former religious authorities were individually accepted as prophets who provided a religious counter to the secular rule of the kings. Late in the 11th century BC the tribes of the north appointed Saul as their king at Gibeah and soon after that the men of Judah in the south of the Israelite area made David their king at Hebron. This was a time of great shifting tribal alliances and splits and to disentangle these in detail would not be beneficial to our purpose, which is to evaluate the Israelites' use of chariots in warfare. When Saul was killed in battle by the Philistines, David fused these two new kingdoms into one kingdom and became its king. He was followed by Solomon, who reigned until 931 BC.

On Solomon's death the Israelite tribes who occupied the north of the Israelite area refused to accept Solomon's son Rehoboam as king over them. Tribal loyalties were still an important factor in Israelite politics, and the northern people declared themselves independent as the Kingdom of Israel. The tribes in the south who remained loyal to the family whom they regarded as the rightful kings of Israel were left as the Kingdom of Judah with their capital in Jerusalem. Both kingdoms were menaced by the encroaching power of the expansionist Assyrians and Israel was captured about 722 BC by either the Assyrian king Shalmaneser V or his successor Sargon II, we are not sure which.

Judah survived numerous Assyrian assaults, including the famous siege of Jerusalem by Sennacherib in 701 BC. Full and immensely valuable accounts of this appear in both the Second Book of Isaiah and the Annals of Sennacherib. Both sides claimed an outright victory in what was in fact the best negotiated outcome either of them could get away with. Hezekiah, King of Judah, was surely the most incompetent of an unimpressive line of monarchs who thought they could safely refuse to pay further tribute to the power with the most efficient army in the Middle East.

There was considerable dislike on the part of the Judeans for the citizens of Israel. In New Testament times its people were referred to by the Judeans as the Samaritans after one of their principal cities and the area in which they lived. After the Assyrians conquered the northern Kingdom of Israel the inhabitants were deported to other parts of the Assyrian Empire and replaced by non-Jews from other places. The point of the parable of the good Samaritan, which must have raised many a chuckle in Judah, was that no Judean could possibly imagine that any Samaritan could be good.

After the collapse of the Assyrian Empire Judah finally succumbed to the Neo-Babylonians in the period known in Jewish history as The Captivity and eventually became a province of the Persian Empire.

The tribe known as the Jebusites appears in the Old Testament. They were that section of the Canaanites who inhabited Jerusalem before the Hebrews invaded Canaan.

David fused these two kingdoms into one. After Saul the term Hebrews for the people was replaced in Hebrew language texts by the term Israelites and they soon found themselves at war with surrounding peoples, the not very militarily powerful Semitic-speaking Moabites and Edomites to the east and the much more powerful and developed Philistines to the south who were attempting to extend their influence northwards. The Old Testament, that collection of theology, poetry, tribal genealogy, and political and military history gives information on at least the outcomes of those conflicts, although not of the manner in which they were fought. The Philistines had a considerable number of chariots, as had the Canaanites who had been squeezed out to the north by the Israelites or absorbed. And so the kings of Israel had to join the arms race and build up an army that replaced the old infantry tribal levy and included a chariot arm.

In the early days of the Hebrew settlement in Canaan they had been virtually restricted to fighting in the mountainous highlands because of the Canaanite strength in chariotry. The Hebrews had been successful when they could lure their opponents' chariots into hilly country where the chariots could not pick up speed and could be attacked by infantry.

This is seen in the encounter in the days before the Israelite monarchy described in the Book of Judges chapter 4, verses 6–17, between the Israelite infantry army commanded by Barak and the chariot army of King Jabin of the Canaanite state of Hazor commanded by Sisera. Sisera had 900 chariots and Barak had 10,000 infantrymen. Barak managed to lure Sisera to follow the Israelite army with his chariots up Mount Tabor where the chariots were overwhelmed by the Israelite infantry. Sisera dismounted from his chariot and escaped on foot. So presumably his chariot had been bogged down or somehow brought to a standstill.

Sisera is remembered in history for the sticky end he came to. While escaping from the battlefield he sought refuge in the tent of Jael, the wife of Heber the Kenite. The Kenites were another branch of the Canaanites who happened to be friendly towards the state of Hazor at that time.

Sisera was thirsty and exhausted and Jael gave him a blanket and a drink of milk. He lay down and slept. Then she took a tent peg and a hammer and hammered the peg into his head with such force that she skewered him to the ground. When Barak turned up she showed him her handiwork with evident satisfaction. Her motivation in the complicated cross currents of tribal hatreds is not explained but she appears to have gone down in Israelite history as a heroine.

Solomon purchased chariot horses from Egypt and from the Neo-Hittite states to the north and the Aramaean states of Syria at 150 shekels each, several times the price of a horse not trained to be driven as one of a pair under a yoke. Chariots were bought from Egypt at 600 shekels each. This was also a fantastically high price. But chariots were the new conclusive weapon, which it was hoped would ensure victory.

The Israelite religious leaders, the prophets, had always warned the people that installing kings as rulers would mean greatly increased taxes. But the ordinary people seemed to be willing to accept the increase in taxation in exchange for the military success.

In the Old Testament both the Second Book of Chronicles and the First Book of Kings, Chapter 4, Verse 26, say that Solomon had 4000 stalls for horses and also had 12,000 horsemen. Much depends on the meaning we give to the Hebrew word that the Authorized Version of the Bible translates as 'stalls' but which could mean teams and the one that is translated as horsemen. The latter could mean men who attended to horses and not men who rode them. For cavalry was still in the future.

If we take the word translated as stalls to mean pens, each containing a single animal, that gives three men to a horse. If we regard it as describing teams, which would mean a pair of horses, 2000 in all, we are left with six men to a chariot. We know that Solomon bought chariots from Egypt and that the very light Egyptian chariot carried a crew of only two men and had not really room for more. But we do not know if the men that Solomon captured included crews who rode in the chariots. Whatever way we manage the statistics, they do not leave the Israelites with many men to hold the heads of extremely excited horses that are raring to go with their fellows.

Another indication of the strength of chariot forces is in the Second Book of Samuel, Chapters 3 to 4. It says that when David defeated Hadadezer of Zobeh and his allies from Damascus he captured a thousand chariots and hamstrung a hundred teams that were surplus to his requirements, in order to put them beyond use.

One gets the impression that the Israelites under David fought as infantry and their tactic was to lure the Canaanite chariotry into hilly country where the Israelite infantry could tackle it. It was certainly after the establishment of the monarchy under Solomon that the Israelites started investing in chariots to a large extent. The chariots were probably mostly of the light Egyptian type but we do not know how they were used tactically in the hilly country for which they were not designed.

The Philistines

Ever since they appeared in the Old Testament as a major enemy of the Hebrews the name Philistine has become synonymous with lack of refinement and deliberate disregard for good taste.

Their best known personage is Goliath of Gath, famous for his size and how easily he was killed with a single sling shot by a simple shepherd boy.

The Philistines are known in Old Testament times principally as the inhabitants of five cities in southern Canaan. These are Gaza, (underneath the modern city), Askalon (modern 'Asqalan), Ashdod (modern Isdud), Ekron (possibly the modern Khirbet el-Muqanna'), and Gath. The sites of the first three are well known. The most likely site of Ekron is thought to be the one given above. No archaeological site has yet been identified as Gath, although there are possible locations.

Our theories for the origins of the Philistines are dependent on the interpretation of Egyptian texts, but this broad outline is generally accepted.

In about 3500 BC various peoples who originated in Anatolia, Cilicia (the part of the Mediterranean coast where it curves from eastwards in Anatolia to southwards in the Levant) and perhaps the northern part of what became known later as Palestine started migrating southwards towards Egypt. The names that survived in Egyptian texts for individual peoples of that group were the Tjekker, the Peleset, the Sherden, the Danuna, the Sheklesh and the Tursha. Collectively they were referred to by the Egyptians as The People of the Sea or The Sea Peoples. Their linguistic origin is unknown and not relevant to our purpose, but it was not Semitic. Some seem to have been taken into Egyptian service as slave troops. Others fought against the Egyptians and some settled in Egypt. They were ejected from Egypt by Ramesses III (1198–1166 BC) of the XXth Dynasty and some of them, named as the Peleset and the Tjekker, moved northwards and about 1200 to 1190 BC settled down in southern Canaan where they became known as the Philistines and were responsible for that area of the Levantine seaboard receiving the name of Palestine. The eastern neighbours of the Philistines, between them and the Dead Sea, were the recently arrived Hebrews and the surviving remnants of the Canaanites whom the Philistines and the Hebrews had displaced.

To followers of the current situation in the Near East I should point out that the Arabs do not feature in this story, although there were nomadic Semitic-speaking tribes called the *erebu* wandering about the Syrian desert at that time. The term *erebu* comes from a Semitic verb meaning to wander. To try to relate ethnic movements in the Near East in the 2nd millennium BC to the current situation there would be a pointless effort.

The Philistines became well established and were seeking to expand their sphere of influence, as were the Egyptians to the south of them and the Hebrews to the east. Maritime trade was a valuable investment worth fighting over.

The first clashes of the Philistines with the Hebrews came from about 1100 BC onwards, with wars becoming endemic between the Philistines and Saul. Neither side won a final victory, although the Philistines were consistently doing better while the Hebrews retained their rule over only the southern and eastern margins of their country. Accounts of Philistine history appear only in Hebrew writings in the books of Samuel in the Old Testament. These are very fragmentary and not concerned with the details of warfare, only the results.

The Philistines clearly had an efficient army and earlier Egyptian reliefs from Medinet Habu of the time of Ramesses III show the Egyptians winning battles with the Sea Peoples. Regrettably we are reduced to hoping that the vehicles of these ancestors of the Philistines were not so different from those of the later established Philistines. The Sea Peoples are easily distinguishable in the reliefs from the Egyptians by the feathered head dresses with a chin strap which they wear.

The Egyptian Medinet Habu reliefs show the women and children of the Sea Peoples caught up in the pursuit after the great battle in which the Sea Peoples were expelled from Egypt. They are seen riding in two-wheeled carts with solid disc wheels and high sides made of various patterns of stout timbers. The axles are under the centres of the bodies of those heavy vehicles that are drawn by oxen and clearly they are not chariots designed for war.

Our knowledge of the Philistine army comes from Egyptian and Hebrew sources. The Egyptian source is the earlier wall reliefs at Medinet Habu and the Hebrew source is the First Book of Samuel, in the Old Testament. Chapter 13, verse 5, is a general description, which says that the Philistine army in the time of Saul included thirty thousand chariots and twenty thousand horsemen in addition to infantry. It is therefore lacking in useful technical information on their chariots and the horsemen can hardly have been organized cavalry at that early time in the 11th century BC. That does not exclude the possibility that mounted men were used as messengers at that time, for we know that horses were ridden then, although twenty thousand of them is surely excessive. But for all that mounted archers existed among the nomads of Central Asia at that time, we cannot accept that organized cavalry was in being in the forces of such a minor power as the Philistines when such an organized major power as Egypt did not have them. And the figure of the Philistine army containing thirty thousand chariots must be treated with equal suspicion. I feel that this, as so often happens in military intelligence estimates, modern as well as ancient, is a deliberate exaggeration of the size of the enemy to increase the importance of any future victory or excuse a future defeat. Battlefield estimates of the strength of the enemy are usually equally exaggerated.

The Canaanites

These were the Semitic-language speaking peoples of the central Levantine coastal region between the sea and the Lebanon Mountains whom the Israelites invaded in their incursion around the late 13th century when they moved into the land that they believed had been promised to them by their god Yahweh during their centuries of wandering in the Wilderness. The Canaanites occupied a string of coastal cities extending from Ugarit, just beyond Latakia, in the north and their territory stretched inland for forty or more kilometres to the River Orontes and south to Gaza and the cities of the plain west of the Dead Sea. One will of course bear in mind that these limits varied from time to time in an area that saw continual wars of conquest and shifting spheres of influence and that no established national boundaries were involved.

We do not know for certain what kind of chariots the Canaanites had at that time, some four hundred years after the manufacture of the Ugarit bowl. There were two sources from which minor countries like the small Canaanite city states could buy chariots with which to build up an army and preserve their independence. These were the Egyptians and the Hittites. Both were willing to supply chariots to suitably useful countries in their wars by proxy with each other. And if the carpenters, joiners, and metal workers of these small countries were skilled enough to build their own chariots, the designs would still probably reflect the preferences of the major powers that were copied. If Egyptian designs were followed these would be light vehicles of lengths of steam bent wood lashed together with animal guts or sinew. Unless, that is, heavier vehicles with metal fittings of Hittite style were copied.

The Old Testament, which was not written by or for military historians, is usually deficient in information on chariot design, concentrating instead on the religious and political aspects of victories. But occasionally a snippet useful to the historian slips through.

One is the statement in Joshua 17:16 in which it is written, 'The Canaanites that dwell in the land of the valley have chariots of iron…' Now these cannot have been chariots made entirely of iron, for they would have been too heavy to move and no such vehicles are known. So they must have been chariots with iron fittings. These could not be Egyptian ones. So that leaves only Hittite chariots. This piece of intelligence seems not to have been spotted by historians before.

It was in 1460 BC that Thutmose III of Egypt led an army that he said contained 1000 chariots, into the Levant to impose Egyptian power over the Semitic-speaking people of the Canaanite city states there. He was opposed by a coalition of Canaanite kings led by the king of the city of Kadesh. Thutmose, whose name is also transliterated asTuthmosis, met up with the Canaanites there. There ensued the battle that I have described in chapter 9. That would have been a most interesting conflict between the super light Egyptian chariots and the heavier Hittite designs.

The chariots of some regions had details that portrayed derivation from different countries and periods. Good examples of these come from the 9th and 8th centuries BC from north Syria, from the cities of Carchemish, Malatya, Tayanat,Tell Halaf, Sinjirli and Til Barsip. Inevitably the foreign influence at those dates in that area was Assyrian and the vehicles show features that belong to Assyrian chariots of earlier periods, although that does not mean they were not still useable.

The Neo-Hittites, sometimes called the Syro-Hittites

These were survivors of the Hittites after the collapse in the 13th century BC of the Hittite Empire in Anatolia who found themselves in small city states in northern Syria where they lived in an amalgam of Hittite culture and the local Aramaean one. The best known Neo-Hittite cities are Tell Halaf (biblical Gozan), Sincerli, Carchemish, Damascus and Malatiya.

As the survivors of a militarized state, the principal profession the Neo-Hittite men knew was that of arms. So they hired themselves out to larger states in Palestine like Israel and Judah as mercenaries in the wars that these states had with major regional powers such as Assyria and Egypt.

The little Syrian states were increasingly under threat from Assyria and found that they could only resist the Assyrians if they allied themselves with Israel. In the end they could not stand up against the might of the Assyrian army for ever and they were eventually conquered and their cities had become provinces of the Assyrian Empire by 700 BC.

The Neo-Hittite mercenary best known to readers of the Old Testament is Uriah the Hittite who is mentioned in the Second Book of Samuel in the Bible. Uriah, who was not in fact a citizen of the now defunct Hittite Empire, was a Neo-Hittite fighting as a mercenary on behalf of the Judeans against an enemy who is not mentioned in the Old Testament. King David of Judah lusted after Uriah's wife, Bathsheba, with whom he was having an adulterous liaison and whom he had got pregnant. So to dispose of the inconvenient husband David arranged for Uriah's troops to withdraw from around him in battle, leaving the Hittite alone to face the enemy. Uriah was killed in action and David got Bathsheba. The lady's feelings in the affair are not recorded. David was criticized by the prophet Nathan.

There is an orthostat of the 9th century from Carchemish that shows a chariot that displays decided Assyrian influence. The horses have high semi-circular crests somewhat like Assyrian chariot horses, but much higher than Assyrian ones and with feathers or hair on top. The wheels have six spokes and have much heavier felloes than the Egyptian ones. The axle is at the rear of the body, which is smaller than Assyrian ones.

There is possible evidence for Neo-Hittite cavalry in a basalt orthostat now in the Berlin Museum, relief number 22, from the palace of Kapara at Tell Halaf. It shows a mounted rider, a cavalryman, with a forward seat. He wears a helmet and carries a round shield and what must be a drawn sword, although because of the volcanic origin of the rough stone it looks more like a truncheon.

The problem is that the palace of Kapara is dated by the excavators to the 10th century BC and the more reliable date for the employment of Assyrian cavalrymen who did not need a horse holder is late in the reign of Ashurnaṣirpal II of Assyrian in the 9th century. But although we can be pretty sure of the dates of that Assyrian king, the dating of the palace of

Kapara to the 10th century cannot be relied upon with that certainty. What it does suggest is that cavalry, mounted men who are capable of fighting together on horseback, was adopted by the Neo-Hittites pretty soon after it had been in Assyria. But it still does not seem to have been taken up until later by the other military powers in Western Asia. Of course the rider seen on the orthostat is not necessarily a cavalryman He could be a solitary messenger.

The small cities of North Syria are a good source for the 8th and 9th centuries BC of pictures of chariots that display foreign features. In the then current political situation dominated by Assyria these could sometimes include features that were archaic in Assyria at that time. These north Syrian towns were Neo-Hittite settlements, although their culture must be taken as being Semitic Aramaean rather than Hittite. In 856 BC Til-Barsip, the capital city of the country of the Aramaean Bit Adini, was captured by Shalmaneser III who had a palace built there. Its walls were adorned with frescoes that contained paintings of chariots that were of accurately depicted dateable Assyrian patterns.

These kinds of chariots are also known from the sites of Sakjegōzu, Tell Halaf, Carchemish, Tayanat, Malatya, Sinjerli (ancient Sam'al, the capital of Yadiya), and Til-Barsip and seem to come in two basic types, earlier ones in the 9th century BC and later ones in the 8th.

They vary in some having the axle under the centre of the body while others have it under the rear. The wheels of the earlier chariots have six spokes while eight-spoked wheels come in at the time when eight spokes also appeared in Assyria after the reign of Tiglath-pileser III (744–727 BC). But wheels with six spokes and those with eight were used in North Syria at the same time. The chariots whose likenesses were found on reliefs at Sinjirli, Tell Halaf and Carchemish have the axle under the centre of the body, while those from Sakjegōzu, Tell Tayanat and in an ivory from Ziwiye have them under the rear. The chariots on the wall frescoes of the Assyrian palace at Til-Barsip have the axles under the rear of the body and wheels that are large and very much after the Assyrian pattern. Two other Assyrian features that appeared on North Syrian chariots were a lance that was placed point uppermost and leaning backwards at an angle from the chariot body and a shield with a lion's face on it mounted on the rear of the body. These features date in Assyria to the time of Shalmaneser III but survived longer in North Syria.

The general impression is gained that there was no standard type of chariot in north Syria at that time but that chariots there were strongly influenced by Assyrian prototypes and aspects of Assyrian chariot design were adopted as it pleased the builders without any slavish adherence to current Assyrian styles.

Cyprus

The evidence for the existence of chariots in Cyprus comes in two forms. These are most importantly in the shape of actual vehicles dating from the 8th and 7th centuries BC and secondly in terra-cotta models. They strongly resemble Assyrian chariots and provide us with information that cannot be obtained from the Assyrian reliefs on the mounting of the shield that is seen blocking the rear of the Assyrian chariots of Ashurnaṣirpal II and Shalmaneser III. An intact Cypriot chariot has a metal loop half a metre high at the rear of the body, which would have been most convenient for holding one of those shields. The floor of the Cypriot chariots is D-shaped and provides us with confirmation that it was made of interlaced thongs similar to those on Egyptian vehicles. They also have the backward leaning spear that is found on pictures of the early Assyrian chariots depicted on the White Obelisk of the time of Tukulti- Ninurta-Ashur.

Chapter Eleven

Anatolia

In this chapter we deal with the three states in Anatolia of whose chariotry we know something and who were deeply involved in the power struggles of the 1st and 2nd millennia BC in the Middle East. They are Hatti, the land of the Hittites, in the 2nd millennium BC and Urartu in the 1st, along with Mitanni, which was a major power for much of that time.

Mitanni is also dealt with here, even though its territory was spread over northern Syria and the north of Assyria rather than Anatolia. But much of its fighting and political manoeuvring involved the Hittites so that politically and strategically it is best lumped together with the states in Anatolia rather than those in Syria to the south.

The Hittites

Anyone interested in learning about the Hittites must be aware that the name covers people speaking three closely related Indo-European languages, Luwian, Palaic, and Nesite. They moved as immigrants into central and eastern Anatolia, presumably from the east, in the Bronze Age about 2000 BC. The area in which these three groups settled was inhabited by people who called it the land of Hatti and themselves Hattians.

The immigrants took over the name Hatti and called themselves Men of Hatti. They adopted the cuneiform form of writing used in Mesopotamia and adapted it to their own Indo-European languages. So, as Mesopotamian cuneiform has been readable by Western enthusiasts since the middle of the 19th century CE and the Hittites are often mentioned in Assyrian records, we know quite a lot about them. The indigenous Hattians were assimilated into the culture of the immigrants, although they left a lasting cultural influence in religion and art.

By about 1700 BC the new Indo-European immigrants into the Land of Hatti in central Anatolia had become the most important group in society. They established their capital at the site of the Hattian city of Hattuš, near the present day village of Boğazköy, but called their own language našili, which is known to modern scholars as Nešite, in contra-distinction to Hattili, the language of the Hattians. They were part of the mass migration westwards in the Bronze Age of people speaking related languages who had their origins in the steppes of south Russian Siberia in the 3rd millennium BC. They spoke languages related to a now unknown ancestor whose language family is known as Indo-European. The oldest known Indo-European language known to present-day scholars is Sanskrit, the liturgical language of the Hindu scriptures. One group of these people, known as the Aryans or Indo-Aryans, moved south into north India.

Here I must clear up a possible point of confusion and mention the recent 20th century CE German National Socialist obsession with Aryans whom the Nazi philosophy identified with the Germans. This is complete nonsense. The Aryans are an early Indian people and the only real Aryans in 20th century Europe were the Gypsies. In spite of this they were persecuted by the Nazis who were looking for convenient minorities to maltreat and because the travellers' peripatetic life style did not fit in with Nazi ideas of the perfect society.

Another branch of Indo-European speakers moved directly to Europe and all the peoples of present-day Europe speak Indo-European languages, except the Basques, Estonians, Finns and Hungarians. Some went west into Anatolia, among them the Hittites.

The Hittite area of governance stretched eastwards as far as the kingdom of Mitanni in eastern Anatolia and to Damascus in the south, where Hatti had a border with the sphere of influence of the other major western Asiatic power, Egypt.

Hatti was surrounded by enemies, Mitanni to the east and Egypt to the south. In between Hatti and Egyptian controlled territory were various small states to the south and west whose names are now of interest only to specialist historians of the region.

So an essential for Hatti's survival was the maintenance of a powerful and efficient army. An important part of the Hittite army was the chariotry. The

Hittite chariot was heavier than that of the Egyptians and designed more as a weapon platform than for speed, although it cannot in practice have been much slower, probably by only a few miles per hour. The six-spoked wheels were of fairly robust construction, perhaps not as hefty as the Assyrian ones but much stouter than the light bent cane wheels of the Egyptians. It is possible to imagine that they were better suited to the rough mountainous country of the Hittite homeland than the Egyptian vehicles which were designed for use in level smooth plains. The axle of a Hittite chariot was under the centre of the floor, not at the rear as were those of the chariots of other advanced major powers. Much is always made in discussions of chariot design of the theory that having the axle at the rear would throw the centre of gravity of the vehicle further forward and make a tighter turn safer. For the principal hazard in a chariot is turning over. But having the axle at the rear places more weight on the horses' shoulders and saddle galls are an ever-present evil. This would not matter in a light one or two-man vehicle like the Egyptian one. But with a Hittite chariot with the axle further forward and a crew that consisted of a driver and up to two warriors clad in armour, more weight would have been kept off the horses' shoulders. Their battle tactics would have been designed to allow for that. We cannot say which design would have been the more effective until such vehicles are used in their hundreds in experiments that are real combats, which are of course impossible. And success in battle depends on many more things than technology.

During the period of the Hittite Empire, from c. 1700 BC to 1190 BC, the term Hittites was applied to the subjects of the Kingdom of Hattuša. After the collapse of the Hittite Empire, some urban centres in north Syria that had been part of the Hittite Empire remained occupied by Hittites who except in language had adopted the Semitic culture of the existing inhabitants. We do not have demographic statistics on the division of the population between Syrian Semites and Anatolian Hittites. But the latter left monumental inscriptions in a developed form of the hieroglyphic script that had been employed by the former indigenous kings of Hattuša. Except for their inscriptions the survivors of the Hittite Empire who lived in Syria seem to have adopted the Semitic culture of the Aramaeans who formed the majority of the population. They are generally known to modern scholars as Syro-Hittites or Neo-Hittites and were discussed in the chapter on Syria. The Israelites of the Old Testament called them Hittites. Ancient peoples, like modern ones, were often confused and inexact as to what they should call foreigners of whom they knew little. The term Syria itself is an anachronism, dating from the Roman Empire. But it is used here because it is the name of a modern state in that locale and is generally understood.

The Hittites were a warlike expansionist people, as were most ethnic groups in the ancient Middle East. We know from Hittite texts that the Hittite king Hattušili (c. 1650–1620 BC) campaigned in Syria and had a long-standing feud with the King of Aleppo. His grandson Muršili I defeated Aleppo and went on eastwards to sack Babylon in about 1595 BC, which brought to an end the Old Babylonian Period. Thereafter with weak kings, which means unaggressive ones, Hittite political and economic fortunes declined. The Mitannians, another Hurrian people with an Indo-European speaking group of rulers, had established the kingdom of Mitanni in northern Mesopotamia and were in the ascendant.

If, like all other armies, the backbone of the Hittite army was the infantry, chariotry was an important shock force. Chariots appeared in the Hittite world in the early 16th century BC, the idea possibly coming from the Mitannians further east.

The Hittite chariot was pulled by two horses. At first it carried two men, the driver and a warrior who had a bow. By the time of the Battle of Kadesh in 1274 BC it had a third man who carried one of the Hittite 'figure of eight' shields and a spear.

Evidence for what a Hittite chariot looked like comes not from Hittite sources but from Egyptian reliefs from Medinet Habu. Their chariots can be identified as Hittite by the figure of eight shaped shield that is carried by the warrior riding in it. It is probable, and not unexpected, that the Hittite chariot and its horses should have items that were copied from the chariot equipment of other countries. The felloes of the wheels were much thicker than Egyptian ones, but not quite as thick as those of Assyrian chariots. The axles in early Hittite chariots were positioned under the middle of the floor, as were Canaanite ones. By the 16th century the axle had been moved to the rear of the floor. And the wheels now had the number of spokes increased from four to six, as had Egyptian chariots.

The main weapon of Hittite chariot crews and their Levantine and Anatolian allies at Kadesh was the spear and there was no indication of a quiver or bow case being carried on Hittite chariots.

But Hittite fortunes improved and the great conquering Hittite king, Suppiluliumas I (c 1380–1346 BC) defeated the Mitannians and made Syria a network of Hittite vassals. Hatti was now a major power in the Middle East.

But now there was a threat from a new direction as Egypt continued with its plan to make Palestine an Egyptian sphere of influence. In 1300 BC the pharaoh Ramesses II led an army, which included four divisions of chariotry, into Asia. He met up with the army of the Hittite Emperor Muwatalli and his Syrian allies and the large battle of Kadesh ensued.

Both sides claimed a victory but in fact it was a stand-off, since it did not change the balance of power. In a treaty in 1284 BC the boundary of Hittite territory under Muwatalli's successor Hattušili III and the stretch controlled by Ramesses II was fixed at a line just south of Damascus. But peace did not last for long. Hittite enemies who were not related to each other combined together for the first time and Hattusas was sacked in c. 1190 BC. It was the end of the Hittite Empire.

Yadin attributed what success the Egyptians had over the Hittites to the Egyptian use of the compound bow rather than the spear as a chariot weapon. Littauer and Crouwel (1979) have demonstrated that thrusting a spear sideways from a moving vehicle is not an effective use of the spear as a weapon. According to the inscription of Pharaoh Mernephtah (1237–1219) at Karnak, Egyptian chariots had a quiver containing eighty arrows.

But these arguments pro and con the spear and the bow as the chariot weapon ignore the psychological impact of charging chariots on the enemies, infantry or chariotry. This does not appear in the accounts of the victories except in the most general terms. No ancient king would want to admit that the enemy whom his chariots charged was already shaken and ready to run. But that is the way it would have been. Chariots charging other chariots head-on can expect equal losses on each side. But when they charge shaken infantry they can turn a defeat into a rout. The classic example of the futility of cavalry charging steady infantry is Marshal Ney's disastrous charge of 10,000 cavalry at Waterloo against unshaken British infantry with fixed bayonets. The French horses would not face the bayonets and the cavalry had to content itself with galloping round the British squares out of reach of them.

A Hittite text of about 1700 BC seems to be the earliest reference to the use of chariots in battle when King Anitta fielded an army that included 40 chariots. Other texts mention the Hittites as having 80 chariots. The number of chariots included in major powers' military inventories increased rapidly. At the Battle of Megiddo in 1482 BC Tuthmosis III captured 894 chariots. By the time of Amenophis II (1450–1425) the Egyptians captured 1092 chariots in campaigns in Syria. In Ramesses II's battle at Kadesh about 1300 BC the number of chariots belonging to the Hittites and their allies and vassals was estimated by the Egyptians at 3700. At the Battle of Qarqar in 853 BC the Assyrian king Shalmaneser III faced a force of 3940 chariots fielded by an alliance of five states. But the same Syrian confederation was said to have 1900 horsemen. If these were cavalry and not just grooms then this was the earliest reference to the use of cavalry in battle.

Hatti was surrounded by enemies, Mitanni to the east and Egypt to the south and various small states to the immediate south and west whose names are now the concern only of professional historians of the region.

The two Bronze Age battles of Megiddo and Kanesh that provide us with details of the tactical use of chariots were between the Egyptians and the Hittites. They are Megiddo in 1482 BC and Kadesh about 1300 BC. They could equally well be regarded as either Egyptian or Hittite battles, although both took place in Syria.

Although Egypt had now acquired chariots its army was substantially an infantry force that had the new, for Egypt, longer range and more powerful composite bow. The Egyptians' opponents consisted of Hittites and their Canaanite and Syrian allies who were stronger in chariots. These had a different kind of vehicle. The Egyptian chariots were extremely light, yet flexible and had a crew of just the driver and one archer. They were designed for scouting ahead of the main army and harassing the enemy without coming to close quarters. The chariots of the Hittites and their allies were heavier, with crews consisting of the driver and two warriors. One was armed with a spear and carried a shield. His duty must have been to protect the driver and both he and the driver wore light textile armour. The other warrior carried a spear and probably later a bow and arrows. He wore a long suit of bronze lamellar scale armour and a helmet. This vehicle was intended to be used to charge to contact and its body was panelled to give the crew some lower body and leg protection, which the Egyptian chariot did not provide. It was, like the Egyptian chariot, drawn by only two horses. For increasing the size of the team of animals beyond two does not increase the speed of the vehicle, while it makes turning more difficult.

Canaan produced no more threats to Egyptian military power for the next twenty years. And there were no more major battles in which chariots played a notable part.

This did not mean that peace reigned in the Middle East. Countries just rebuilt their strength for the next war. Tuthmosis III now had to turn his attention to curbing the power of Mitanni. But there were no major battles.

The Hittites under their expansionist ruler Suppiluliumas I were steadily building up their power. Mitanni was in terminal decline. Egypt was in no state to involve itself deeply in foreign affairs as it was in the throes of the internal dissention caused by the attempts of the pharaoh Akhenaten (1379–1362) to replace the traditional Egyptian religion with its powerful and well-entrenched priesthood with a new religion of his own choosing, the worship of the physical disc of the sun, the Aten.

With the disappearance of Akhenaten's disruptive efforts to alter the even tenor of Egyptian society by introducing a new strange religion of his own it was Ramesses II (1304–1237) the third pharaoh of the XIXth Dynasty who next set out to come to terms with the Hittites and establish Egyptian influence in Syria. His father, and predecessor on the throne, Seti I, had fought the Hittites, including a battle at Kadesh. Yet although the resulting peace treaty delineated spheres of influence, no firm boundaries were laid down and the Egyptians must have let the Hittites keep Kadesh in their area.

But the next big battle the Hittites had to fight was about a hundred and eighty years after Megiddo was round the town of Kadesh, some three hundred kilometres north of Megiddo. There are differences of opinion about the exact date of the battle of Kadesh, which in some publications is given as 1274 BC, while others restrict themselves to c. 1300. The name of the town is also sometimes spelled Qadesh with a Q instead of a K, because of the transliteration into the Latin alphabet from the Arabic one of two different Arabic letters that although written very differently in both the Arabic and Latin alphabets have closely similar pronunciations in Arabic.

The battle took place in the course of another of those trade wars that the Egyptians, in this instance under the pharaoh Ramesses II, fought in an effort to control the valuable trade in luxuries from the east. The Hittites, under their king Muwatallis, were concerned to keep the Egyptians out of their sphere of influence. The detailed account we have of the battle comes from Egyptian sources such as the bold and dramatic depiction of the battle the temple at Luxor. Afterwards the Egyptians claimed a great victory in what in fact was more of a drawn affair.

The Egyptians celebrated and memorialized Kadesh as a great victory in a hieroglyphic inscription and a pictorial wall relief. But the final peace treaty signed eighteen years later and sealed by Ramesses' marriage to Muwatallis' daughter was in effect a mutual defence pact and a demarcation of the Egyptian and Hittite territories in Syria. It lasted until the collapse of the Hittite Empire a hundred years later.

But the Egyptians did achieve political influence over a large stretch of Syria that had previously been part of the Hittite sphere of influence and the battle showed that properly handled chariots could be extremely successful in surprise attacks on infantry that was not prepared to receive them. It also showed, although whether this lesson was learned at the time is another question, that chariotry, like the later cavalry, is a one-shot weapon than can be used only once. We know from later wars that cavalry horses will charge, but many will never do it again.

After a slow start during the reigns of a succession of weak kings in the 17th and 16th centuries the Hittites were fortunate in having strong rulers with expansionist ambitions in Tudhaliya IV and his son Suppiluliumas in the 14th century and they eventually destroyed the kingdom of Mitanni.

The final result of the encounter at Kadesh was inconclusive but the Hittites won the peace and managed to push the limits of their sphere of influence south as far as Damascus. Meanwhile Assyria was a rising major power south-east of the Hittites in Mesopotamia.

But no empire lasts for ever. Eventually it becomes too big to govern or it runs out of wealth or a supply of young men who are willing to risk their lives for the benefits in prestige or land they might acquire from being members of a victorious army. Armies that are continually on the defensive after the previously inferior enemies have learned the power of forming coalitions do not have the will to win of those that are conquering valuable national assets.

The immediate causes of the collapse of the Hittite Empire were attacks by tribes from the north and internal upheavals, and after its fatal attack by enemies who were not related to each other and had never combined before Hattusas was sacked in around 1190 BC and the land of Hatti split up into smaller tribal kingdoms. It was the end of the Hittite Empire.

Mitanni

Mitanni was a state in southern Anatolia and north-west Mesopotamia that rose to prominence in the 16th century BC and grew to become a major Middle Eastern chariot-owning power in the then current vacuum in the regional struggle for influence. It had a Hurrian general population ruled by an Indo-European speaking aristocracy, judging by the Indo-European names of the gods they worshipped. The members of this ruling group were famous as chariot owners and skilful chariot drivers and were known as the *maryannu*, which is taken to mean 'noble chariot warrior'. They were also connected with land holding and the status might be hereditary.

Mitannian chariots were not greatly different from those found in other countries in Western Asia. They were lightweight, tending more towards the Egyptian designs than to any others. They might have had four, six, or eight spokes on their light wheels that were possibly copied from Egyptian originals. The axle was under the rear of the body. Where they differed from those of other countries was in the protection afforded to the horses. They had a covering made from felt or hair three centimes thick and some of them in addition had scale armour over the top of that.

The Middle East was enjoying one of its reasonably quiet war-free periods. The Old Babylonian period was coming to an end; Assyria had not yet risen to the position of dominance it later enjoyed; the Old Hittite Kingdom was crumbling as a result of palace revolutions and in Egypt the early pharaohs of the New Kingdom were busy consolidating their power rather than engaging in foreign wars. Mitanni took full advantage of this relatively quiet situation to expand its political influence over Syria and

Assyria. The Assyrians, who called it Hanigalbat, became its vassal. So when Egypt started its VIIIth Dynasty forward policy in the Levant, Mitanni was its principal opponent.

The Mitannian homeland was in the land between the northern reaches of the Tigris and Euphrates rivers, south-west of Lake Van and on a latitude with Lake Urmia to the east. But at the extent of its power it expanded considerably and controlled many of the city states of Syria that had formerly been vassals of the Hittites. Its capital was the city of Washukkani, which has not yet been discovered but is thought to lie somewhere in the region of the Khabur River in northern Mesopotamia.

Unfortunately Mitanni left us no written records so what we know of it comes from Assyrian, Hittite, and Syrian texts and we do not have technical details of the *maryannu* chariots from Mitannian sources. But we can be pretty certain that they cannot have differed substantially from the medium weight chariots of the Hittites. They had probable four to six spokes in the wheels and had the axles at the rear of the body. So they were a pretty conventional type with no outstanding regional peculiarities. Fortunately we have an archive of tablets written in Akkadian cuneiform from the Hurrian city of Nuzi, thirteen kilometres south-west of modern Kirkuk in northern Iraq. The Nuzi tablets provide us with lists of breeds of horses and records of the purchase of horses from breeders.

We have no picture of a Mitannian chariot. Stillman and Tallis (1984) illustrate one that is just an Egyptian chariot of the reign of Tuthmosis IV, but to my mind we have no proof that the Mitannians used this kind of light chariot. In their mountains a heavier Hittite-style chariot would have been more robust.

In the 2nd millennium, round about 1300 BC, Mitanni was the dominant power on an area of southern Anatolia and Northern Syria that stretched in the east as far as Assyria, in the south to the Mediterranean coast beyond Ugarit and the Orontes River and in Anatolia over the whole south-eastern quarter of that land mass. To the west of them in Anatolia were the Hittites while to their south along the Palestinian coast were numerous small Semitic Amorite states in what later constituted what became known as Phoenicia.

In the jumble of peoples in that area with no concept of nationality with the exception of those under a few major powers with strong government like Egypt the division the historian has to make between different peoples has to be a linguistic one.

Looking at the chariots in the Middle East on a regional basis one is struck by the similarity in the trend in chariot design there and in the Aegean in the 2nd millennium BC. In spite of design differences between countries, they shared a common basic design that marked them as very different from the chariots of eastern Asia. The Canaanites, from whom the Egyptians drew their inspiration in chariot design, favoured the extremely light vehicle. Although deficient in protection for the crew it had the advantage of lightness, which increased both the distance the two horses could pull it in a day's work and its speed in battle. This was, presumably, at the expense of its stability. The Assyrians preferred the heavy chariot drawn first by two and then later by four horses. Its higher sides gave slightly more protection for the crew than did Egyptian chariots but the missile power that either chariot carried, two archers, did not differ.

In the Second World War the tank design preferences of the two countries that possessed large tank armies, Soviet Russia and Germany, followed divergent paths. The guns of the two countries' tanks did not differ significantly but the Soviets went in for large numbers of faster more lightly armoured and cheaper to build tanks while the German designers produced heavier, slower, and more thickly armoured vehicles. The important difference between chariots and tanks is in the motive power. Heavier tanks demand bigger and more powerful engines. The motive power of all chariots remained at two horses until towards the end of the Assyrian Empire in the reign of Ashurbanipal when the number of horses was increased to four, which provided more traction in hilly country.

But Mitannian ascendency did not last for more than two hundred years and following on a series of palace revolutions and royal usurpations it was conquered by the newly ascending Hittite Empire further west in Anatolia while Ashur-uballit I of Assyria (1365–1330) shook off Mitannian overlordship and set Assyria on the road to empire. Thereafter Mitanni disappeared from history.

Urartu

The Urartians were a Hurrian people who established a state in the first millennium BC in eastern Anatolia round Lake Van and in the region of the River Zab. Their territory was east of the land of the Hittites. By the time that the Urartians had established a stable state around the end of the 9th century BC, the Hittite Empire had long gone. Their territory was highly mountainous, with ranges running east and west and separated by comparatively narrow valleys. They were regarded as a threat by the Assyrians because of the ease with which they could menace and control the trade routes that ran east and west across the northernmost parts of the Assyrian sphere of influence in Mesopotamia and in north Syria. So for the Assyrians they were a menace to be controlled by military means.

Not that diplomacy was lacking in the ancient Middle East. Evidence of it comes to us, if rarely in formal treaties, more often in the announcements of the marriages between members of the royal families of countries that had competing interests. But as Clausewitz (2008) said, war is policy by other means and the Assyrian kings were religiously bound to lead a campaign against someone each year. So if tribute collecting was proceeding quietly in the west and the mountain dwellers in the east were not pressing too strongly against the Assyrian frontier, that year's enemy might as well be Urartu.

Ancient Middle Eastern wars were at heart about trade and control of the routes along which the valuable and prestigious goods were carried. If you could build a garrisoned town in a valley along which went a trade route you could make a good living from charging caravans customs dues to pass. We know from the extensive archives of the Assyrian kings how much they valued exotic things, even plants and animals from far off countries, for the way in which they enhanced the kings' prestige in the eyes of their subjects.

In 1270 BC Shalmaneser I of Assyria mounted a campaign against a group of eight societies in the Lake Van and River Zab regions. These would not necessarily include the Urartians. Their day had yet to come. But it shows that even this early in the rise of Assyrian power eastern Anatolia was regarded as a threat.

Tiglath-pileser I of Assyria (1115–1077) led a campaign to the north and defeated sixty kings of Nairi. Nairi was the one of the Assyrian names for Urartu and 'kings' could mean little more than tribal chieftains.

Shalmaneser III (858–824) led three campaigns into the Lake Van region in country that would be not at all good going for chariots. By the end of the 9th century Urartu had become so powerful that it was a threat to Assyrian prosperity because of its control of the trade routes in north Syria.

But Urartu did not last for long. Possibly already weakened by the attacks of the Assyrians, among whom the king Tiglath-pileser III (744–727 BC) was the most successful in crushing Urartu's Syrian allies, Urartu collapsed in the late 8th century under attack by the Medes. The flourishing and highly developed Urartian state had lasted no longer than a hundred and fifty years, having fallen prey to yet another group of Indo-European horse-breeding nomads from further east on the Iranian plateau. It was the Medes who, allied to the Babylonians, sacked Nineveh and brought about the fall of the Assyrian Empire in 612 BC in what is accepted as the convenient date for a politically messy business. The Medes themselves succumbed to the Scythians, another group of hard riding horse nomads from out of the east, opening yet another chapter in the continuing saga of ethnic incursions into western Asia and Eastern Europe.

The Urartians had well made chariots, although in what numbers we do not know. They were drawn by pairs of horses no doubt provided by the Medes east of Lake Urmia who also supplied high quality animals to the Assyrians. Evidence for the chariots of Urartu and what they were like comes in an unprovenanced bronze plaque now in the Metropolitan Museum of Art, in New York. It dates from the reign of one of the kings of Urartu called Argishti. We do not know which one, but it is believed on the balance of probabilities that it was Argishti II (714–c. 685 BC). The two Urartian chariots on it closely resemble the heavy Assyrian ones with the axle under the rear of the heavy panelled bodies, wheels with eight spokes and thick felloes. These chariots are clearly in processional scenes, not war or sporting ones, and the crews consist of the driver and one other character. They both wear pointed helmets but are not in any other apparent armour. They are not using weapons. The chariots have the sloping lances or standards projecting up at an angle from the rear of the body that were a north Syrian feature that was so typical of the early Assyrian chariots of Ashurnaṣirpal II (883–859 BC).

A repoussé likeness of another Urartian chariot is to be seen on a bronze helmet whose picture was published in a Russian publication of 1828 by F. E.

Schulz. It also is based on the Assyrian pattern and has two lines running from the end of the chariot pole to the top of the body, similar to that which is interpreted as being a subsidiary support for the pole in Assyrian chariots.

How many chariots the kings of Urartu possessed is unknown to us or how they were used in battle. They must have been of limited, if of any, use in the severely mountainous terrain of eastern Anatolia and were probably restricted to use in the valleys. It is possible that like more recent modern battleships, they were held in the military inventory because every important country had them.

Chapter Twelve

Europe: The Mycenaeans

The first appearance of wheeled vehicles in pre-historic Europe was in the 2nd half of the 3rd millennium.

A theory that has attracted much attention is one that holds that between 4400 and 2800 BC migratory waves of steppe pastoralists swept across prehistoric Europe bringing with them the domesticated and tamed horse and wheeled transport. This is not now accepted without reservations about the plausibility of much of the supporting cultural evidence. But it is an attractive theory.

There is difficulty when excavating the remaining two wheels of a vehicle in deciding whether the burial is of a two-wheeled cart or part of a four-wheeled wagon. If it is decided that it is a cart, that does not necessarily mean that it is a chariot, which is a military or a processional vehicle.

Calibrated C_{14} dates would place the Earlier Bronze Age at c. 2500–c. 1500 BC, which would exclude Mycenae. Mycenae is the first appearance of light, horse-drawn, spoked-wheel vehicles between the Orient and the Aegean.

Before the end of the Earlier Bronze Age came a technological revolt with the development of the domesticated horse as a traction animal in association with a light vehicle with a pair of spoked wheels.

The Mycenaean period in Greek history takes its name from the well excavated archaeological site of Mycenae ninety kilometres south-west of Athens. The Mycenaean age of Greek history that lasted from the 16th to the 12th centuries BC has given its name to the Late Bronze and Helladic periods of Greek history. Those covered the time of the Trojan War of which Homer wrote about in the *Iliad*, although this epic poem was not in fact written until around the 8th century, three hundred years after the end of Mycenaean power. So while Homer's tale is set in a Mycenaean context, it has a mixture of Mycenaean background and the customs and habits of his own time and he clearly did not know a lot about chariots.

Culturally, Mycenaean influence extended beyond the Greek mainland across the Aegean to the island of Crete, where the very advanced Minoan culture that the Mycenaeans greatly admired was in full flower.

The Minoan culture had started earlier, about 3000 BC. Written texts of the Minoan culture from the 18th century BC survive. But their script, known as Linear A, has not yet been transliterated or translated. A later script, Linear B, which dates from the 14th century BC was transliterated and translated in a linguistic *tour de force* by the late Michael Ventris and has been identified as an early form of Greek.

That was a period of warfare in which the chariot played an important part. War was endemic in Mycenaean times with a considerable number of small Greek states such as Achaea, Pylos, Thebes, Tiryns and Knossos in Crete of closely equal power jockeying for political supremacy. The chariot was a Mycenaean development that spread from mainland Greece to the Minoans in Crete. When Minoan culture collapsed about 1450 BC, probably because of an earthquake, Knossos was taken over by the Mycenaeans and Mycenaean culture became the dominant one there and throughout the Greek sphere of influence. The Mycenaeans developed trading links with the Hittites, Egypt and Syria.

The personnel of the Late Bronze Age Mycenaean armies was composed of spearmen, swordsmen, lightly armed skirmishers and crews of heavy chariots that probably joined in fighting their set piece battles on what restricted level plains there were and in which chariots could operate. The unsprung chariot needs enough space in which to make turns and pull up from manoeuvres at speed and the ever-present danger to its occupants is that it may turn over, bringing the terrified horses down with it.

Greece lacked the expanses of wide plains that Syria possessed in which the extremely light chariots of the Egyptians could safely pick up speed and, most important for a chariot driver, space in which they could pull up once the enemy force had been passed. There are no brakes for chariots.

So the Mycenaeans built heavier chariots than the Egyptians did. Apart from the team of two horses galloping side by side at twenty miles an hour towards an enemy force they depended for their military threat on the warrior with his long spear standing beside the driver. How effective he was as an aggressive threat is a matter of opinion. If he stabbed with it without letting go of the weapon then it was no more than local protection to deal with any enemy infantryman or charioteer who was rash enough to venture within an arm's length of a fast moving vehicle that could not stop. If he had used it as a javelin and thrown it, then we know from our tests with a replica Sumerian battle wagon that he could hit with reasonable certainty any person or object within several metres. But for that he would have needed, like the Sumerians, a quiverful of javelins ready to hand and there is no evidence that the Mycenaean chariot warrior had that reserve of weapons. Certainly the Greeks did not pack their chariots with two or three archers in addition to the driver as the Assyrians did.

But we do not know how much these long spears were for show. They would certainly have been impressive. I would contend that the threat of a chariot was principally the appearance of it rushing at already shaken infantry. And also in the Bronze Age for a country to be taken seriously as an important power it had to have a great many chariots. A similar aspect of national pride and the psychology of preparation for warfare can be seen in the need of 19th century South American republics to own at least one battleship, whatever the cost and no matter how slight was the likelihood of an imminent war.

But this is a long-standing problem with the military use of chariots to which we have not an answer in the present day. For economic reasons modern controlled and measured tests to replicate the performance of chariots have always been done with single vehicles, and inevitably have excluded actual warlike tests in which men are put in danger or killed.

It could be that the actual, in fact the only, aggressive power of a chariot was the psychological effect of the approach on the enemy of the team of galloping horses bearing down on them with the same foot poundage as an artillery shell. A massed charge of them must have been a frightening prospect. But it would only have taken one less herd conscious or more neurotic charging horse to run out for there to be an almighty pile up.

The Assyrian chariot archers and the Mycenaean spearman can only have been there for local defence against any enemy brave, or foolhardy, enough to attempt to attack the chariot on foot from the side. To win, the enemy who was being charged had two choices, stand firm and scare off the chariot charge with long spears or else, if he had room enough and there were not too many chariots around, open up his ranks and let the chariots thunder off through the gap into the distance behind him. Of course a wise chariotry commander, like any wise cavalry one, did not order his men to charge unbroken infantry. The same goes for infantry bayonet charges.

In practice anyone close enough to a moving chariot to attack the crew would already have been run down by the horses, so the spearman would only have been useful once the chariot was brought to a halt and had no other close support. Of course this is theoretical supposition and without the practical experiments, which are unfeasible, we cannot be at all sure of the outcome. In the end that would vary according to circumstances and the motivation of the humans involved.

Drews (1993) suggests, convincingly, that it was long swords and infantry with throwing javelins that spelled the doom of the great chariot armies of the Late Bronze Age. I would add that in spite of its poor manoeuvrability it was its speed that was the chariot's superior asset. Once the skills of riding had improved the same job could be done more easily, and more cheaply, by cavalry. For a ridden horse is more agile and able to get out of trouble than a horse yoked to another. In effect, riding a horse is safer than driving one if you get into trouble. An old soldier pays more attention to staying alive than to winning a gallantry award.

There exist no pictures of a Mycenaean chariot battle. But although the Mycenaeans armed their chariot warriors with spears and not bows, Minoan Linear B tablets from Knossos show that the Mycenaeans possessed and knew how to make compound bows.

The pull on foot pounds, which is easily measured, required to start a light military horse drawn vehicle and then keep it in motion is so little for two or more horses that the main task for the driver is not to keep his animals galloping alongside other galloping animals but to keep the vehicle from turning over as it rocks, leaps and bumps over the slightest stone or irregularity on the ground.

This all comes down to in my opinion, based on driving and riding experience, that a chariot is an extremely dangerous vehicle to be in unless the people you are charging are already scattered and running away.

Once the chariot, a light two-wheeled vehicle with spoked wheels pulled by two or more horses under a yoke was invented about the 16th century BC, it spread rapidly throughout the Levant and the Middle East. Although it is possible to find minor improvements to it that can be traced to a certain culture, it is not practicable to attempt to say who invented it first. Like 19th century battleships or 20th century tanks, every country that aspired to be taken for a major power had to have one, hundreds if possible. I am suggesting that like a battleship it was more impressive than useful.

Three different kinds of Mycenaean chariots are known.

The Box Chariot was in use between approximately 1550 and 1450 BC. The evidence for it is an engraving on a signet ring found in a shaft grave in Mycenae. It had a square body, presumable open at the rear, with wheels with four spokes on an axle under the centre of the floor. It was pulled by two horses of presumably the early European breed standing thirteen hands, 52 inches (1.33 metres) in height at the withers, now the height of a larger child's pony.

The body, which came up to upper thigh height was panelled in some material such as wickerwork. The body was attached to the yoke by a conventional chariot pole with an S-bend in it that was attached to the front of the base of the body. The pole had an additional support of a light rod, which ran from the top of the front of the body to the yoke. This had thinner vertical stays that hung at intervals from it down to the main chariot pole and their ends were lashed round it. This contraption is now considered to be an equivalent support and shock absorber to the elliptical device that is found on Assyrian chariots of the reigns of Ashurnaṣirpal II (883–859 BC) and Tiglath-pileser III (724–727 BC). But it is questionable whether it would have had the ability of the Assyrian device to even out the shocks to the main pole that might break it away from the body. The crew was a driver and an armoured warrior wearing the typical Mycenaean helmet made of boars' teeth laced together and an armoured corselet whose exact composition could have varied. It could have been an articulated bronze corselet made of wide overlapping bronze bands several inches wide of which three were wrapped round the wearer's trunk to cover him from hips to chest like the one that was found at Dendra, near the Mycenaean citadel of Midea. With it the wearer's shoulders were protected by shaped rounded plates that covered the shoulders, overlapped in the centre of the chest. The arms were left bare and the armour extended up to protect the neck and possibly also the lower face. There is evidence from other finds that this kind of armour was not uncommon. From this it is clear that the Mycenaeans possessed considerable ability in metalworking. The excavators dated it to about 1400 BC.

The warrior wielded a long spear that would have to be used with two hands to keep attackers at a distance. Archers appear in engravings of chariots only when the vehicles seem to have being used for hunting, not in warfare, a departure from the Egyptian or Assyrian practice.

The Dual Chariot that was in use from about 1450 to 1200 BC was a modification of the Box Chariot that had two semi-circular panels fastened to the rear of the sides of the body. Their purpose has been the subject of speculation. They could have been hand rails to assist in mounting the vehicle or additional guards to protect the crew from flying stones. The floor of the vehicle was rounded at the front.

A rare type of Mycenaean chariot known only from one or two carved representations is termed the Quadrant Chariot. It appears to have been used only for a short period between 1350 and 1375 BC and had only one occupant, the driver. It had a D-shaped floor like the Dual Chariot but its purpose is unknown. Perhaps it was an experimental model.

The last type of Mycenaean chariot was the Frame Chariot, which was used from about 1250 to 1150 BC. This was a super lightweight model whose openwork body was made of bent canes. It probably had the axle under the rear of the floor, not the middle.

There is no direct evidence as to how Mycenaean chariots were used in battle.

Until the 1970s it was accepted that Mycenaean riders fought as cavalry. It is now thought by historians that the Mycenaeans rode to the battlefield and then dismounted and fought as infantry.

There is a complete lack of archaeological evidence for chariots in Greece between the 12th century BC when pictures of them appear on Late Helladic IIIC pots and the 8th century when they appear on geometric pottery and bronze and terracotta figurines.

In a few decades before or after 1200 BC there was a dramatic change in the weapons and armour used in the eastern Mediterranean. The infantry corselet appeared as did the Naue II bronze sword with the grip cast along with the blade. This did not appear in the Assyrian Empire where swords remained thin with complicated ornamental grips that can hardly have been cast in one piece with the blades. But we are relying on reliefs, not actual weapons. Armour in the Late Bronze Age was simple. Chariot crews were reasonably well protected with a robe reaching to the calf and covered in mail. Then in the Aegean a fully protective bronze suit of armour dated to the 15th century was found in a chamber tomb at Dendra. The Dendra corslet encased the wearer from the neck to the knees. It would have so restricted movement that it is thought that it must have been for a chariot warrior.

Drews suggests that in the Aegean in the 12th century new light spear heads about 11 centimetres long appeared in Greece and the town of Ugarit on the Syrian mainland. They were lighter than those of stabbing spears but too large for arrowheads. He considers these to be throwing javelins and sees them as being the great antidote to chariots and that they brought about the end of chariot warfare in the Aegean. But this was the beginning of the Assyrian Empire with its strong chariot arm, which goes to suggest that the development of chariotry followed a different path in Mesopotamia from that of Greece and Syria.

Archaeological authorities have considered at length the possible relationship between the spoked wheel and the disc one. In his chapter on the earlier Bronze Age in his book *The Earliest Wheeled Transport* Piggott (1983) discusses the relationship between the adoption and development of the two-wheeled light vehicle in the 2nd millennium in the Near East and Europe. He goes along with Gordon Childe's 1954 suggestion that the spoked wheel was a new invention rather than an adaptation of the tripartite disc. Littauer and Crouwel in 1979 suggested the possibility of a local evolution of the light spoked wheel horse-drawn chariot in the Near East, in contrast to a theory that had long been held, that the spoked wheel was introduced from outside by Indo-European steppe tribes. They hold the view that Mesopotamian peoples were familiar with horses from early in the 2nd millennium BC and the linguistic and lexical evidence used to support the theory of an Indo-European origin of the chariot is too late to be relevant. They see the spread of the chariot into Europe on the other hand as the result of diffusion from higher to lower cultural levels.

Piggott sees war chariots and chariot warfare as specialized products of social situations that did not exist in Europe in the 2nd millennium BC. He does not favour the view that many held before he wrote his ground-breaking study that the idea of the light two-wheeled horse-drawn chariot spread into central and western Europe from Mycenaean Greece alone. Instead he considers that it entered Europe from the Levant, Egypt, Asia Minor and Mycenaean Greece together.

In Europe the traditional use of disc-wheeled carts and wagons extended from the 3rd millennium down to and beyond the Roman period. The bulk of the evidence for vehicles in Europe from about 1300 to 700 BC is in the form of bronze attachments and sheaths for the spokes of vehicles that must have been used for cult or prestige purposes.

In Greece there is no archaeological evidence for wheeled vehicles between about 1200 BC and the first appearance of chariots and funeral carriages in paintings on Geometric vases in the 2nd third of the 5th century BC. Two alternative explanations present themselves. Either there is evidence of continuity from Mycenaean times that has not yet been discovered, or else vehicles were abandoned before being reintroduced from the Near East. Jost Crouwel (1978) opts for continuity, assuming a military function for some of the two-wheeled vehicles seen painted on Geometric vases of the second third of the 8th century.

Others prefer to think that chariots were reintroduced from outside for racing and processional purposes.

The Iron Age of the 7th and 6th centuries in central Europe has produced the remains of what are taken to be four-wheeled wagons. But if they might have been in fact two-wheeled vehicles they had been dismantled and it was impossible to tell if they were chariots. Surviving pictures of vehicles are too sketchy for us to be sure what they were. One drawing on a pot shard from Rabensburg in Austria shows a two-wheeled vehicle with eight spokes and a triangular body with a 'stick man' standing driving two animals under a yoke. But we do not have accompanying social evidence to lead us to believe it was used in war. So the evidence for Early Iron Age chariots in Europe is inconclusive.

Chapter Thirteen

China

An unexpected aspect of Chinese history is that such an advanced and innovative society should be so far behind Western Asia in the acquisition of the wheeled chariot. And in fact these do not appear in Chinese archaeological sites until around at least five hundred years after they are found in the West.

China had not been shut off from Central Asia where the domestication and taming of the horse took place. And knowledge of that technological advance had percolated through to Western Asia possibly eight hundred years before it appeared in China. And the spoked wheel, whose adoption is regarded by scholars as essential to the building of the fast chariot, is found at Siberian sites that are dated to at least the 2nd millennium BC.

It is surprising that a society such as China, which later saw the invention of the stirrup that made the heavy armoured cavalry of mediaeval Europe possible and whose people were leaders in the invention of the mariners' compass and gunpowder, should have to wait until a chariot with numerous spokes to its wheels was produced in Armenia in the mountainous terrain of Central Asia before they took it up. In the Shang Dynasty (c.1766–1045 BC) the Chinese had to fight enough against the nomads in the West to be aware of such an invention as the chariot that was already well established in the West.

It was not until 1200 BC or possibly a century earlier that the horse was taken into use as a working animal in China, if indeed the animal itself was known there. And there is no evidence of a wheeled chariot until around that time.

However, in considering the position of the chariot in Chinese warfare we are presented with Chinese Bronze Age texts with more extensive instructions on strategy than we have with Near Eastern kings' annals. In fact what we have are thoughts on what the commander's attitude towards war ought to be, rather than the all too common Near Eastern quasi-religious political justifications for the campaign and panegyrics on the power and military skills of the king.

Yet with the Chinese not only do we have detailed information in the form of the remains of decomposed wood of the dimensions of the unusual Chinese chariots, but also texts that are philosophical treatises on the moral attitudes the successful general ought to have on the strategy of warfare.

The nearest we have to these from the West are various much later Byzantine texts on strategy and tactics. But they are of a strictly practical nature, unlike the philosophical Chinese ones.

There are six principal Chinese texts on military strategy so far discovered. Where there is a generally accepted English title, it is used here. Otherwise the Chinese title is given, transliterated into the Latin alphabet. The names and dates of the Chinese texts in approximate chronological order are:

Text	Presumed Author (where known)	Date
Six Secret Teachings	T'ai Kung	Warring States period 11th century BC
The Art of War	Sun Tzu	6th century BC
Military Methods	Sun Pin	Middle of the Warring States period c. 300 BC
Wei Liao-tzu	Wei Liao-tzu?	4th century BC?
Wu-tzu	Wu ch'i	4th century BC
Questions and Replies	Huang Shih-kung	7th century CE

The texts were found in tombs, written on strips of bamboo that were tied in bundles. While they give us a valuable window into the workings of the Chinese mind, they are of limited usefulness to the military historian trying to work out how the Bronze Age Chinese used the chariot in battle. They are more concerned with the spiritual qualities the general should possess. That is the reason why in the Orient they are studied to this day as guides to success in business. But they do show that chariots were used with subtlety and imagination. Some examples that deal with chariots are:

Use chariots to mount incendiary attacks on supplies under transport.

Military Methods

In defending ditches and moats, chariots are used as fortifications.

Military Methods

In unknown enemy territory send out your infantry ten miles ahead and your chariots and cavalry a hundred miles ahead.

Six Secret Teachings

When crossing a river by a pontoon bridge use the Martial Assault chariots at the front and rear. Use them to block off all the intersecting roads and entrances to the valley.

Six Secret Teachings

Therefore in chariot fighting, when ten or more chariots have been taken, those should be rewarded who took the first. Our own flags should be substituted for those of the enemy and the chariots mingled and used in conjunction with ours. The captive soldiers should be kindly treated and kept.

The Art of War

When there is dust rising in a high column, it is the sign of chariots advancing.

The Art of War

When the light chariots come out first and take up a position on the wings, it is a sign that the enemy is forming for battle.

The Art of War

In the operations of war, where there are in the field a thousand swift chariots, use as many heavy chariots …

The Art of War

We do not know the significance of Sun Tzu's distinction between swift and heavy chariots. For only one type of Chinese chariot that would seem to merit the title of swift rather than heavy has been excavated. But this awakes the archaeologist to the possibility that another kind of Chinese chariot existed.

The Art of War is the best known of the Chinese military texts, and also the most useful for our present purposes. For it holds out tantalizing possibilities of further discoveries in the field of chariot history that are waiting to be revealed.

The Chinese military texts are in general rather later in date than what we can learn from Middle Eastern texts on the use of chariots in their area. But it has been suggested that at least one of the Chinese texts harked back to principles that had been laid down in the 10th century BC.

After the Neolithic the first Chinese dynasty of the Bronze Age was the Hsia. It is generally reckoned to have lasted from 2100 BC to about 1600 BC, when it was overwhelmed by the Shang Dynasty. The Hsia society was basically a Stone Age one, and although bronze working of a fairly basic standard had started, its products were confined to the upper classes. There were numerous wars during the period but they were fought by infantry and there do not appear to have been any chariots involved.

The Shang Dynasty was the first from which useful historical documents survive. It is characterized by advanced bronze casting and the appearance of wheeled vehicles. Chariots first appear in China in the reign of the Shang king Wu Ting about the 13th century BC. The history of the Shang Dynasty in the middle of the 2nd millennium BC is one of turmoil and wars against smaller competing states. The Shang were, however, generally victorious, although operations in mountainous terrain and in very hot areas were difficult for armies raised in the plains. Chariots were of extremely limited usefulness without the bows with which the crews could launch arrows at high velocities over longer ranges than the opposing javelins could be thrown.

Here I would record my dislike of the anachronistic term 'fire' that is so often used in connection with bows and arrows. Arrows are launched, released, loosed or shot. But there is no fire involved. That applies only to firearms.

The Shang Dynasty is a most important one for our study because it was in its reign that the chariot was adopted for its numerous wars against neighbouring peoples. But the origins of those enemies are unclear and their names are of more interest to sinologists than to specialists in horse transport. Excavation has now shown that Shang power extended over all of Henan Province, parts of Shanxi Province, most of Hubei Province and parts of two others in the valley of the Yellow River. Whether this was a purely a military occupation of foreigners or the extension of direct rule over local chieftains who were already nominally Shang vassals is not clear. But the area includes several important ancient cities that show considerable Shang influences.

Excavations by Chinese archaeologists in the 1980s have uncovered an important Shang city by the Luo River, near the Yellow River in Henan Province, whose ancient identity is not yet certain. It is dated to around 1600 BC and wheel ruts were discovered on an interior road. They are thought not to have belonged to war chariots.

History starts in China and prehistory ends with the introduction of writing and advanced bronze casting in the Shang state in the southern coastal region of the country with its capital at Yin on the Hwang Ho river in about 1500 BC. Its power did not extend beyond its military centres and it was not yet a centralized state. The history of China in the Bronze Age is one of competing military aristocracies, all ethnically Chinese, whose power over the mass of the Stone Age population was based on their possession of the new strategic alloy, bronze, and a monopoly over its casting. Their politics were driven by the complicated loyalties of ten clans as they warred with each other. Chinese history in this period is turbulent and violent, although probably not more so than that of the Middle East with its struggles for power between expanding states and its continual invasions from the east by different ethnic groups. West of the Shang state were the people, also ethnically Chinese, known as the Chou, who were sandwiched between the Shang and the nomads of the steppe. They had better made chariots than the Shang and more horses.

Chariots were not unique to the Shang during the Shang Dynasty. A Chou oracular inscription mentions the capture by the Shang late in its period of two chariots. Ongoing archaeological exploration on the periphery has revealed well-constructed chariots in Shandong that belonged to a smaller regional state and are mentioned in texts appearing in late Shang contexts.

The horse was already domesticated and tamed and it is considered possible that the Chou may have had chariots before the Shang. What we do know is that the Shang had acquired two-horse carts, which might be military chariots, by 1300 BC.

The Shang dynasty was overthrown by the Chou by about 1227 BC. That in turn collapsed after military defeat about 707 BC and was replaced by the Ch'in Dynasty in 256 BC. It lasted only until 202 BC, when the different warring states of China were united by the Han dynasty.

The first scientific excavations in China, between 1929 and 1937 were at Anyang in the south of northern China in the modern province of Henan. They revealed a particularly advanced culture of the Shang period. Among the objects found were the soil traces of the remains of carts or chariots. Knowledge gained from these has been added to by similar remains found in more recent Chinese excavations. These vehicles had survived as buried traces of decomposed wood and it is a tribute to the skill of the excavators that it was possible to reconstruct the form of the vehicles from those soil traces.

One of the advantages of the discovery of buried traces of the Shang vehicles is that we have their dimensions, something we can only guess at for the Assyrian chariots. The dimensions of the Shang vehicles, working from a published drawing of a typical chariot and ignoring the engineering maxim that you should never take the dimensions of a part from a drawing, are:

Wheel diameter	1.5 metres
Number of spokes	8 or more
Body length back to front	1 metre
Body width, side to side	Approximately 1.4 metres
Wheel track	2 metres
Height of sides of body	Approximately 0.4 metre
Pole length, from body to yoke	2 metres

The wheels are much larger than Middle Eastern chariot wheels and while the weight of these vehicles can only be guessed at, they would have been much lighter than an Assyrian chariot, if slightly heavier than an Egyptian one. The fixed axle was under the centre of the body.

In a Chinese chariot with its low sides the crew would at speed, I guess, have had their work cut out to stay in the vehicle even if they were kneeling.

This leads us to a big question for which the archaeologist would dearly like to find an answer. With no drawings of the Shang cart or chariot to guide us, what was it used for? Was it a fighting vehicle or only a command post? Or was it restricted to being a processional vehicle for the aristocratic notables in whose tombs their traces have been found? The academic controversy is very much alive. But the various Chinese texts on war leave us with the strong supposition that in China chariots were used in battle.

A Chinese vehicle of this type was found that had the remains of a tower on it.

This suggests that it was an observation point for a commander, not a fighting vehicle. But was it unique?

One thing that comes immediately to mind with the Shang dynasty vehicle is its similarity to the cart whose remains were found at Lchashen in the Republic of Armenia. The Lchashen one was of much the same date as the Chinese vehicle.

Both have large wheels with many spokes, a wide track, and an unusual body that is wider than it is long and both have a low fence of spars no more than about half a metre high round three sides of the body. A difference between them is that the Chinese vehicle lacks a fence in the rear, which is what one would expect in a war chariot, while in the Lchashen one the lack of a fence is at the front, facing the horses. That would be a serious disadvantage in a fighting vehicle.

Lchashen is close to the 40th parallel of latitude half way between the easternmost point of the Black Sea and the mid-point of the western shore of the Caspian. The route goes first south-east to skirt the southern end of the Caspian Sea and thereafter turns roughly eastwards to China. It is 6500 kilometres across rough country, through high mountain passes, from Lchashen to Anyang, which is near the coast eight hundred kilometres south of Peking and there was an established trade route in silk and spices from Central Asia to China that for most of the way crossed country that was above 900 metres above sea level. It passed Tashkent, split to pass north and south of the cold Taklamakan desert, which is of an oval shape 1000 kilometres long from east to west and 400 from north to south at its widest, and then eastwards into China. There is some evidence that the local climate was warmer in antiquity than it is nowadays but it was still a very long walk.

But we have no evidence of any sort as to the purposes for which these vehicles were employed. They could be fighting vehicles, if hazardous ones to ride in under enemy missiles, or commanders' transport and vantage points and prestigious processional vehicles. Certainly they are less severely functional than most war chariots.

They could have been regarded as being of native Chinese design had not a similar unusual design been found in Armenia three and a half thousand miles to the west of north China. The Lchashen vehicles have already been described by me in chapter 1 on Central Asia and their dating is too close to that of the Chinese vehicles, which are thought to date from around 1200 BC for us to estimate from the date alone whether the design was Central Asiatic or Chinese in origin. Dating to a part of a certain millennium BC is as precisely as we can ascribe a date in that period in east Asia with any confidence.

It would be probably be at least twice as far if you had to walk from Armenia to China than it would be if you could have gone in a straight line. But time went slowly in Asia in the Bronze Age and people there are still great travellers, on foot if they are nomads. And it would be possible to walk from Lchashen to Yin in a year, which would be a year well spent if you had highly saleable intelligence on you. Lchashen is in the Republic of Armenia, which at the time of excavation was the Armenian Soviet Socialist Republic, and the remains of the Chinese vehicles were excavated in the People's Republic of China, which at the time of excavation was the Republic of China. Countries in that part of the world have often changed their official names with changes of government and ethos.

Now the archaeologist has to decide whether he thinks that the idea of this peculiar design of wheeled cart originated in Armenia or in China. Or he could think that it might have been invented independently and simultaneously in the Bronze Age in both countries. I think that last option is improbable. I grant that many fundamental ideas such as the wheel or the use of horses to draw a vehicle could well have occurred independently to people in very different cultures. But these vehicles under discussion here with their large wheels with numerous spokes and low openwork railings round the bodies are so different from chariots in any other part of the world and so basically like each other that they must be related.

So, did this highly unusual design of vehicle originate in the poor societies of the mountains of Armenia or the rather richer plains of China? You have to decide whether you think it more probable that it went from Armenia, a poor country that was bordered by China or the other way round, from richer to poorer. Nowadays richer countries often provide armaments to poorer countries. That is so that they can indulge in a war by proxy with another rich country that advances the first rich country's power struggle in the area. But that idea will not wash in the Asiatic Bronze Age. Although the Chinese certainly had military problems with the nomads to the west, Armenia was too far away to be involved. If someone had got hold of details of a revolutionary new piece of military technology, would he try to sell it to a poorer country or to a richer one that was known to have military problems but had no tradition of horse drawn vehicles? This is only a guess, but an educated one, when I suggest that the idea for these light weight manoeuvrable vehicles originated in Armenia and someone calculated that he would be handsomely rewarded if he sold the idea to the nearest rich world power. That would be the Chinese. The Chou capital might be thousands of miles away but he could walk there with his valuable intelligence in a few months. It would be well worth the trouble.

We know that Shang armies were small, averaging 3000 to 5000 men, although many more could be raised in an emergency. In their battles Shang weapons were the compound bow, spears, and a typically Chinese weapon, the dagger axe. That had a bronze blade like a dagger mounted at right angles on a shaft three to six feet long. Lamellar armour in the early Shang period was made of non-metallic plates tied together, but later on bronze scales were used.

This kind of cart or chariot appeared about 1300 BC at the time of the foundation of the Shang capital at Yin, but we believe from the tombs of wealthy people in which they were found that they were restricted to the aristocracy. They were few in number, particularly among the poorer people to the west. A 12th century list of booty taken by the Shang in a campaign in the west mentions 1570 prisoners, fifteen pieces of armour, but only two chariots. Shang chariots, presumably of the type described above, were drawn by two horses in the early part of the period, although four were usual later on.

The crew of a chariot, assuming that such a vehicle was used in action, were the driver, an archer, and a third man who was armed with a spear or halberd or a dagger axe. We know from texts that the archer was stationed on the driver's right hand side and the warrior on his left. The vehicles were said in texts to be supported by archers on foot.

The bows could be made of silkwood thorn, privet, mulberry, orange wood, quince, thorn and bamboo, with horn and sinew glued and bound with thread on it. Bamboo was the least desirable material for bows, even although laminated bamboo bows were common.

Shang armies also included elephants of the Indian species in their entourage. But there is no evidence that these were driven by mounted mahouts Scanty textual records suggest that they were stampeded towards the enemy lines. But hunting reduced the population of wild elephants in China and they were not used for long. Also, elephants panic easily and are a large target for missile weapons. So they have never proved reliable additions to military equipment.

The Shang had two kinds of shields. This must have included those of the chariot crews. They had a small one 70 by 80 cm (27.5 by 31.5 inches) in size with a vertical handle in the middle and which was slightly outwardly bent. They were made of leather with fibrous material underneath and a wooden frame. Later versions were painted with red, white, yellow and black patterns and tigers and dragons. High-ranking officials had bronze plates on them. There were also long ones that could protect a driver. The first metal helmets appear in the Shang Dynasty.

The power of the Chou people who succeeded the Shang lasted from 1027 to 771 BC when their capital, Hao, which was further up the Hwang Ho river from Yin, fell to the Ch'in from the west, although Chou survived as a minor state until 256 BC.

Chou armies introduced improvements in weapons like the dagger axe and also in chariots, although these latter retained the same general form. Wheels now had more spokes, up to twenty-eight, and these were dished outwards to give more strength. Chariots were now pulled by four horses who had rectangular bibs of lamellar armour hanging down over their chests and had their harness decorated with cowrie shells. As time went by, and this happens with all military vehicles throughout the ages, the chariots were burdened with more accessories, in this case with bronze bells and ornaments. They must have tended to become top heavy. Bronze swords with blades about 18 inches (0.5 metre) long came into use. A Chou Chinese text gives instructions on the types of armour that should be manufactured. One was a sleeveless vest of scales of rhinoceros or buffalo hide and the other was leather scales on a fabric backing. In addition to their lamellar bibs, chariot horses now had tiger skins across their backs.

Chou battle accounts emphasize the use of archers shooting from chariots, so they were not solely prestige or command vehicles. A text does mention a tower on a chariot from which a commander could view the battlefield better. So perhaps these very exposed vehicles had a multitude of uses. In them protection for the crew was sacrificed in the interests of all round unobstructed ability to shoot arrows.

If the nobility riding in chariots were the backbone of Chou armies, the common people who fought as infantry were also important. According to a Chou period text, *The Book of Songs*, the peasants had to undergo a month's military training each year. And according to accounts in another text, *The Book of Documents*, the Chou army at the battle of Mu, in which they defeated the Shang army, was instructed to advance slowly and halt at intervals to form up in a straight line abreast. This sounds like very deliberate infantry tactics.

With the Shang the chariots seemed to have operated together in units of 100 or more, rather than dispersed at will.

Some excavated Shang chariots have unusually large wheels 1.6 metres in diameter. The axles are 3.00 metres long and the rectangular body is rounded on the left front. It is said to be 0.34 metres high. Presumably this is the height of the guard rails. The body is 1.17 metres wide at the front and 1.34 metres wide at the back, where there is a narrow opening and a depth of 1.02 metres It is not clear what this last measurement means. Perhaps it is the front to back measurement. There was a sacrificial victim and two horses.

Chariots became the symbol of power and existed as a threat from the speed with which they could be brought into action even though they held only one archer. Powerful states retained thousands of chariots.

Chou chariots became more specialized, with more rugged vehicles being built for transportation rather than fighting. In the Warring States period that followed the Chou period, chariots began to be used as lookout towers, battering rams, moveable ladders, and to carry crossbows. Chinese chariots had much larger wheels, up to 75 to 90 centimetres, than Middle Eastern ones. They were dished towards the centre rather than towards the outside, as is usual in the West.

Shang chariots averaged 16 to 20 spokes. The individual spokes were 3 to 4.50 cm in diameter and wheels were wood with no bronze reinforcements and hubs were long and thick, 20–25 cm long and 18–28 cm thick. In contrast, the hubs of the apparent precursors at Sintashta and Lake Sevan were much thicker at 40–45 cm.

The felloes of Chinese chariots were of two or three pieces of wood softened and bent. They were generally thicker than wider, often 10 × 7.5 cm and there were also some square versions of 8 × 8 cm. No tyres or studs have been found.

The average size of the floor of Shang chariots was about 138 × 96 cm with the vast majority being 130 to 145 cm wide. Small ones of 94 × 75 cm were found and chariots of 150 by 90 and 170 by 110 cm have been found.

The frame enclosing the crew and keeping it from falling out was of rattan, cane, wood, or bamboo 40–45 cm high and extended below the rim of the wheel alongside them. Two variant frames as low as 22 and 30 cm have been excavated and occasionally the rear frame is higher at 50 cm than the 30 cm high front frame. In a few cases a crossbar was found placed above the sides of the frames and as a hand hold at the front.

The chariot appeared out of the Altai Mountains 1900–1800 BC and was adopted by the Shang 500 years later. This was despite the probable introduction of the horse into north-west China about 2000 BC. Meanwhile the chariot was spreading through Central Asia, including the area around Lake Sevan, about 1600 BC south-westwards into the Near East and south-east into China.

In spite of assertions by some scholars to the contrary, there is no evidence of horses being ridden until the Spring and Autumn periods and Warring States periods of the Chou Dynasty when cavalry was created to counter the threat of nomad invasions from the steppes. Patriotic national desires for a country to be regarded as the originator of innovations rather than importer of them must be guarded against, especially where they concern publications from the People's Republic of China.

The chariots would seem to have charged in a dispersed manner but with lines of five. They were ten paces apart, with forty paces between each line of five and the lines twenty paces behind each other. This allowed the infantry to act as close protection and avoided collisions and crashes.

The evidence for this is in a text on strategy of the Warring States period after the Shang Dynasty, but it is reasonable to assume that this was established Chinese doctrine in the Shang Dynasty. We have the advantage of knowing more about the tactical use of chariots in China than we do about them in the Near East.

It is thought that the limited numbers of chariots used by the Shang means that they were used as command vehicles and archery platforms and that the massive numbers mentioned in Chinese texts were relevant to the time after the Shang period.

Swamps were a hazard, as indeed they still are with modern tanks. Chariots were suited to dry plains. A Chinese text of the 6th century BC gives valuable advice on what terrain is suitable for chariots and the hazards of the ground that should be avoided. Such strategic advice does not appear in the West until Byzantine times. In China we lack the pictorial evidence for the use of chariots that we have in Assyrian reliefs but we have much more logistical and tactical advice on their employment than we have from the Middle East. For those with an interest in the practical handling of chariots the two different sources complement each other.

Chinese texts give advice on the use of chariots in action and the enemy moves that are to be avoided. The rails round the body are so low that it is probable that the crew drove and fought kneeling rather than standing. There would not have been room for four men in the crew.

The axles' higher placement than those of Middle Eastern chariots would have resulted in a high centre of gravity and made lightweight vehicles even more unstable than heavier and lower ones. Experiments driving two and four-wheeled vehicles at speed have shown that stability was a constant major problem.

Many of the texts dealing with the handling of chariots in action date from the Spring and Autumn periods that ended about 479 BC and the Warring States period that followed it. But we can be pretty safe in thinking that the same tactical and logistical problems had been encountered in the Shang Dynasty 900 years before.

The Chou army kept the technology of the Shang with a few modifications. The dagger axe now had a haft of up to 18 feet in length and was used two-handed. This gave greater reach when used from a chariot. The chariot wheels were now larger with more spokes and were dished outwards from the hub to the rim. The horses, of which there now usually four, were decorated with bronze frontlets and strings of cowrie shells. The vehicles were now covered with bronze bells and ornaments. It has been said that these additions suggest that the chariot was now used more as a shock weapon than for its mobility. It was well-known in the 19th century of our era that light cavalry always turned into heavy cavalry in peace time as more and more equipment and weapons were loaded onto the horse just in case it was needed. The same could well have been true of chariots in the Chinese Bronze Age. We certainly see the same process with Assyrian chariots.

Much of what we know of the tactical use of chariots in China comes from the texts of the Chou Dynasty but we can apply their recommendations to the Shang Dynasty with care and some reservation. We know that a large army of the Western Chou could contain 3000 chariots and 30,000 infantry. Chariots operated in units of 25 with each vehicle protected by 25 infantrymen.

During the Eastern Chou period the hundreds of independent states were reduced to eight. Chinese aristocratic society, which controlled the privately owned chariots, was built up of mutually antagonistic clans.

The Eastern Chou Dynasty was dominant in 770–256 BC. The states on the periphery of the area of influence of the eight principal Chinese states were expanding out against barbarian peoples. There were considerable cultural differences between the states but they were becoming centralized by a civil service based on merit. A state's power was judged by the number of chariots that it could field and this was in general expanding. For example the state of Tsin, north of the Hwang Ho river had 700 chariots in 632 BC and 4900 by 537 BC, while Ch'I, east of Tsin had 100 in 720 BC but over 4000 by the end of the 5th century BC. It was an age of endemic warfare as increasingly powerful states competed for power. Major wars were between the north and the south, fighting for land to support increasing populations.

Chariots of the Shang dynasty and Eastern Chou periods date to c.1200 to 800 BC and are uniform with only small dimensional differences.

Chariot bodies are rectangular with proportions between length and depth between 4:3 and 3:2. The width is usually between 1 and 1.5 m and the depth 0.75 and 1 metres. The chariot body is usually open at the rear with a uniform height of 40–45 cm. The axle is in the middle of the body and is about 3 metres long. The wheels are about 1.4 metres in diameter and have 18–26 spokes.

It would be helpful here to summarize the similarities between Chinese and Lchashen chariots and the principal differences between them and Middle Eastern ones.

1. The number of spokes in the wheels of Chinese and Lchashen chariots is higher than the four, six or eight found in the Middle East.
2. The Chinese and Lchashen chariots have a centrally placed axle, while those in the Middle East have, after the early part on the 2nd millennium BC, the axle placed under the rear of the body.
3. The felloes of the wheels of the Chinese and Lchashen chariots are of two pieces of wood bent using head, rather than of several pieces of wood carved as part of a circle.
4. The bodies of the Chinese and Lchashen chariots are wider than they are long and have low openwork rails round them while the Middle Eastern chariots tend to be longer than they are wide and have slightly higher panelled walls at the fronts and the sides.

The military role of the Chinese chariots is suggested by three of the chariot pits which contained human skeletons that were armed.

Chariots in the Western Chou period appear to have been more for military purposes than for prestige. The conclusion is that the chariots at Anyang in the Shang dynasty were more for royal status as parade or command vehicles. In the western Chou dynasty chariot remains become more numerous and are unmistakably military.

Piggott (1983) suggests that when an earlier form of the chariot is discovered found in an area in which its later fully developed form is found we must presume that an even earlier heavier and more clumsy form of transport before the chariot, of which we have as yet no archaeological knowledge, must have existed there as well as also the earlier domestication of the horse. No earlier wheels or wheeled vehicles dating from before their complicated Shang Dynasty chariots have been found in China. The horse is not native to China and does not appear in China until it is associated with a chariot in the late Shang dynasty.

A point of some debate in recent years has been the military effectiveness of the chariot. Scholars incline to one of two views.

One is that the chariot was a radically revolutionary weapon that speeded up combat and made possible the circling outflanking attacks that bring success in battle. Once that basic military need had been satisfied, their possession became an increasing source of prestige for states that aimed to become world powers.

The other is that chariots were limited in their usefulness, being of military advantage only in plains. And in the Near East towards the end of the second millennium major powers were being threatened by men from the hills against whom chariots were ineffective because they were not as capable as cavalry in crossing rough ground. But the great powers retained chariots because the possession of great numbers of them marked a state out as a great power that was not to be meddled with. There are more recent modern parallels.

Historical evidence for the use of chariots in China comes from three sources.

1. Oracle bone inscriptions of the Shang dynasty.
2. Bronze inscriptions of the Western Chou dynasty.
3. Historical texts, especially of the Spring and Autumn period.

It is the few extensive inscriptions on Shang dynasty oracle bones used for divination that tell us something about the Chinese use of chariots. A group of these relate to the king's use of a chariot in the hunt for rhinoceros, wild horses and deer.

The two oracle bones whose inscriptions relate to the use of the chariot in battle tell how the western enemies of the Shang dynasty in the state of Gongfang used chariots. So when the intelligence of the new weapon travelled from Armenia to China it was used by the states through which it passed on the way for the Shang to build some for themselves. Such knowledge would be too good not to be put to immediate use.

There exists an early 2nd millennium BC account written by a Shang vassal on a cattle scapula, which tells how two chariots of a western enemy of the Shang dynasty were captured.

Although buried chariots have been excavated in Shang contexts, there is no literary evidence of Shang use of chariots in battle, while we know that they were used by the enemies that the Shang troops overcame. But military history is full of examples of obsolete hardware and equipment that was not originally intended for employment in action being pressed into use because of the 'contingencies of the service'. So if the Shang possessed chariots and such equipment was being employed by Shang's enemies, I feel justified in suggesting that the Shang used their chariots in battle in addition to whatever original role they had for them.

Of interest to us is the dearth of evidence for the use of chariots by states such as Wu and Yue that were enemies of the Shang dynasty in the Spring and Autumn period (722–484 BC). But their territories were marshy and crisscrossed by streams. Chariots could not have operated there.

The numbers of chariots that are listed as having been captured in Chinese campaigns of the 2nd millennium BC are under a hundred while the infantry prisoners were in the thousands. So it is unlikely that China ever saw the massive chariot armies that took part in battles in the Near East. But when reading of these numbers we must be aware that the ideogram that is taken as meaning chariots could equally well have been used for agricultural carts. But we know from archaeological experience that cultures that used farm carts often produced little clay models of them and these do not appear in China.

But it appears as if once again the possession of more and more chariots was becoming a prestigious sign of importance. For in the Spring and Autumn period the state in one area boasted that it could produce 4000 chariots for a review and it is said to have possessed, in all, 4900 of them.

Even the small state of Zhu had 600 chariots. But this growth of the large numbers of chariots in an army was at the time of the demise of chariots as fighting vehicles. For by the end of the Spring and Autumn period mounted horse warriors were already appearing on the north-western borders of China. The era of the massed chariot armies came just before the chariot itself became obsolete. This was to happen again in military history. The perfection of plate armour that made a man almost impervious to blows came just as the invention of artillery and the hand gun, itself a small cannon, made the horseman in plate armour obsolete. The battleship with its long-range 15-inch guns reached its state of perfection as two new weapons, the submarine and the torpedo, made it obsolete. The chariot was made obsolete by a new weapon, the cavalryman.

I end this review of the chariot in China by summarizing my conclusions. The chariot was introduced into China about 1200 BC when it was used in small numbers and was probably a military command vehicle. After the Western Chou dynasty succeeded the Shang, chariots became much more numerous and were definitely used in battle. By the end of the Western Chou period Chinese warfare had become one of massed chariot battles. Shaughnessy (1988) suggests that the possession of a number of chariots by the Chou may have been a factor in their defeat of the Shang, even although the Shang had many more infantrymen than had the Chou. An estimate of 170,000 or even 700,000 Shang chariots, if one can believe those numbers, did them no good against the even larger masses of Chou fighting vehicles.

The ridden horse did not appear in China till the 4th century BC. In this China lagged behind the Near East.

The construction of the wheels of chariots differs significantly between China and the Near East. In the Near East there are four spokes in earlier times. Later these increased to six or eight. In China wheels are lightly made, with between twenty-five and thirty spokes, and the maximum number found was forty-four. The later Iron Age four-wheeled carriage found in Barrow 5 at Pasyryk in the High Altai of Siberia of the 4th or 5th centuries

BC has 1.6 metres diameter wheels with thirty-four spokes and the felloe is two-piece. This contrasts with Lchashen, where the chariot wheels have around twenty-eight. The bending of felloes in antiquity is attested to in both the *Iliad* and the *Rig-Veda* but known in the West only in later Celtic chariots. As Kossack hinted, the constructional techniques that Caucasian and Chinese chariots have in common are too similar to be coincidental or the result of parallel development.

As to horse breeding, it seems to have been established along the Silk Road as far east as Ferghana in modern Uzbekistan and the mountainous Tien Shan area north of the Taklamakan desert on the borders of China and Kyrgyzstan by early in the 1st millennium BC. This was half way between the Caucasus and ancient China and only 3200 kilometres from the borders of the Shang empire. It was in Ferghana that the later emperors of China sought the famous 'Heavenly Horses' that excelled as chariot horses in the 2nd century BC. Piggott (1974) suggests that the same area might have been of interest to the Chinese as a source of horses as early as the 2nd millennium.

Lu Liancheng has published in the journal *Antiquity*, vol 67, 1993, a most informative article on chariot and horse burials in ancient China. They are in Shang and Chou sites at Anyang in Henan and Xi'an in Shanxi. The burials belong to a burial system that lasted for a thousand years from the Shang dynasty (c. 1300 to c. 1050 BC) down to the Qin-Han period (221 BC to 220 CE). They are in pits around the burials of important aristocratic personages and the horses and their drivers were slaughtered in order to serve their masters in the after-life.

The discovery of the burial pits started in the 1930s at the site of Yinxu near the present day city of Anyang and since then further excavations have revealed several hundred chariot burials in pits in Henan, Shaanxi, Shanxi, Shandong Hebel, Inner Mongolia and Hubei province. It is important to note that the chariot appears in China in a fully developed form, with no earlier versions in existence. Only minor changes to it were made after the Shang dynasty. While similar to each other over a large expanse of land and a long period of time, they differed greatly from the chariots found in Egypt, the Near East, Anatolia and the Aegean in the shape and design of their bodies and the large number of spokes in their large wheels.

Below the yoke in Shang and Western Chou chariots hung two yoke saddles. These were found associated with Egyptian chariots but not with those of other Near Eastern countries. Most Chinese chariot bodies were rectangular. The railings round them were on average 30–40 cm high. Some had an extra railing above the normal ones, which suggests that the crew were meant to stand on those chariots. The floors of the chariot bodies were either of boards covered in reeds or else inter-woven leather straps. The wheel felloes were often of just two sections and these were fastened to each other by bronze clasps. Shang period wheels usually had eighteen spokes while those of the Chou period had from eighteen to twenty-six and those of the Spring and Autumn Period (8th to 9th century BC) had between twenty-five and twenty-eight. The chariots excavated from near the tomb of the first Emperor of Qin (c. 221 BC) had thirty spokes.

A feature of Chinese chariots was the very long hubs of the wheels, 20–35 cm in the Shang dynasty but longer in Western Chou times.

Lu Liancheng gives detailed dimensions of a total of twenty-seven Chinese chariots, 5 from the Shang dynasty, 8 from the Western Chou Dynasty, 6 from the Spring and Autumn period and 8 from the Warring States period. From eight selected aristocratic burials he gives a total of at least 61 chariots and the remains of 106 horses.

He is insistent that the chariots were used principally for war. Most of the Shang chariots were driven by two horses. Then, in the Spring and Autumn period (722–481 BC), four-horse chariots appear. A highly detailed bronze model of a chariot and its team from the burial of the First Emperor of Qin has it drawn by four horses abreast. The innermost pair is harnessed to yoke trees that hang under the yoke. The inner extremity of each yoke tree has fastened to it a trace that leads back to a bronze hook in the draught pole forward of the chariot body and from there back to the axle. These traces must have served as additional strengthening to keep the yoke from rotating away from its proper position at a right angle to the draught pole. The outermost pair of horses are harnessed to loops round their chests, which are attached via leather traces to the chariot body. They have no fixed connection to the yoke.

The main developments in chariot design between the 14th and the 3rd centuries BC were an increase in the length of the draught poles, an increase in the number of spokes and a shortening of the axles with a corresponding narrowing of the track of the wheels.

The Chou capital, Hao, fell to Jung barbarians in 771 BC and after a rebellion inside the Chou state in 707 BC it fragmented into a number of small states. This is a convenient spot at which to close this survey of early Chinese chariotry as it was not until 307 BC, well outside our self-imposed time scale, that China's first cavalry units were formed in the Spring and Autumn and Warring States periods.

Chapter Fourteen

The Indian Sub-continent

It is as equally unexpected as the late introduction of the chariot into China that the Indian sub-continent should produce no evidence for the chariot in the period 3000 BC to 600 BC with which we are concerned. For there is no evidence that the Indus Valley Culture or the Indo-European immigrants who swamped it somewhere around 1700 BC had chariots before the end of the 1st millennium BC.

One would have expected the chariot to be taken up as a valuable addition to the armoury of the immigrants. And they came from Central Asia, which was the natural home of the horse. But we have no signs of the chariot in early texts, or as models, or in pictures. Maybe their immigration into the land of the Indus Valley Culture was totally peaceful and not a military invasion. Perhaps they did not need chariots because the earlier indigenous inhabitants did not have them. But that never stopped an invading movement taking up a spectacular and impressive addition to its display of military power, even if there was no immediate military counter to it.

But the fact remains, India and points east, although equally beset with wars, were along way behind Western Asia in adopting the chariot.

The term Indian sub-continent has a certain 19th century ring about it but I use it because it avoids the present division of that land mass into three political entities along religious lines as India, Pakistan and Bangladesh. For that is a recent one, dating from British decolonization in 1947 CE when the sub-continent was divided into two according to the religious loyalties of the Hindu and Moslem inhabitants. The Hindu parts became India and the Moslem parts Pakistan. That was divided further into two parts, Pakistan and Bangladesh, separated from each other by 993 miles of India. The eastern part of Pakistan, Bengal, separated from the western part, Pakistan, in 1971 and became the People's Republic of Bangladesh.

India and Pakistan are intensely antipathetic to each other. In fact they have long been at war with each other over the state of Kashmir which is in India but has a Moslem population. The Moslem religion dates from the 7th century CE, so is not within the timescale of this study. So we are left with 'Indian sub-continent' as the best politically neutral term for the area.

The well excavated early part is in the west, now in Pakistan, in the valley of the River Indus, and its 3rd and early 2nd millennium BC architecturally advanced sites are ascribed by archaeologists to what is called the Indus Valley Civilization. It has two large and extensively excavated city sites, Harappa and Mohenjo-Daro, which date from between c. 2600 and 1900 BC. Texts survive from this culture but the language has not yet been translated. Numerous attempts have been made to relate the Indus Valley Civilization to Mesopotamia, to the Sumerians. But although there were doubtless trading links between the two countries, cultural influences do not appear to have gone beyond these. And the languages of the two regions are not related.

The known sites of the Indus Valley Civilization extend away from the valley of the Indus River eastwards along the coast of the Gulf of Campay. That is from modern Pakistan into India and also northwards up the various tributaries of the Indus. They are found along the Ravi, the Sutlej, and the tributaries of these rivers. They are also found to the west, in the border lands where Pakistan meets Afghanistan and Iran.

Mohenjo-daro is beside the River Indus in Sind while Harappa is beside the former course of the River Ravi, a tributary of a tributary of the Indus, four hundred miles to the north-east in the Punjab. Both it and Harappa were large cities, three miles in circumference and the sites bear all the signs of an evolved civilization with their massive halls, granaries, and dwellings of high officials.

In Mohenjo-daro they had, to judge by a clay model found there, farm carts similar to those of the modern peasant with two solid wheels, which were probably pulled by oxen. The carts have socket holes in their bases, which could support upright spars that would restrain farm produce like hay. But there is no evidence that they had chariots. No models or pictorial representations of them have been excavated and until the Indus Valley language is translated we cannot say if they ever in their eight centuries were involved in war with expansionist neighbours or it remained a unique happy country that had no history.

But the idyll did not last, although its demise might have been a peaceful one. About 1700 or 1800 BC newcomers speaking an Indo-European language moved south from possibly central Asia east of the Caspian Sea into the Indian sub-continent replacing the Harappan Indus Valley culture. They brought with them the caste system and their own Vedic religion, which developed into the Hindu religion. Its sacred texts are a series of hymns or Vedas, of which the principal and best known one is the Rig Veda.

Wheeled vehicles, feature prominently in the Rig Veda, although some authorities suggest that the references to what are translated as chariots may be later than the main part of the text. The military chariot owed its speed of around twenty miles an hour, an increase over the twelve miles an hour of the solid wheeled Sumerian battle wagon or the two miles an hour of the solid wheeled ox-drawn farm cart or wagon, to the lightness and strength of its spokes. These were a Central Asiatic invention that was well known in the Near East by the time the Indus Valley Civilization collapsed around 1300 BC. But there is no firm evidence that the knowledge had penetrated as far south as India at that time. We do not know why. Little solid wheels are found associated with models of farm vehicles from Harappan Indus Valley archaeological sites. Some of them have radiating painted or inscribed lines on them and some scholars have suggested that these are the representations of spokes. But the evidence to support this suggestion is not convincing, for if this means that the vital invention had reached India, the idea was never pursued further.

There is no evidence that the Indo-European successors to the Indus Valley Civilization made any use of military chariots. Portrayals of wheeled cars for the gods and modern-built copies of them play a considerable part in certain Hindu festivals but those who would derive from these a historical connection with Indian chariots in antiquity are ignoring the lack of archaeological or historical evidence. This does not justify us in assuming, on that basis, that chariots were used in the Indian sub-continent within the time scale of this study.

No survey of mounted warfare in the Indian sub-continent would be complete without a mention of polo. It is often said to have been invented as a form of training for cavalrymen. For information on polo and its predecessor, buzkashi, I am indebted to Captain Robert Thompson, an authority on buzkashi.

The origins of polo are in buzkashi, the mounted game played with the carcass of a goat, since long ago and still to this day by members of the Turki-speaking tribes of the Central Asian republics. Today each tribe has its own rules for playing it, but the best known version is that in which two mounted teams attempt to drag the decapitated carcass of a goat tied up in a skin bag through the opposite team's goal. By some accounts it goes back to the 6th century BC while others favour the beginning of the current era. But there is no linguistic, pictorial, or literary evidence for those dates.

What is more certain is that the Mongols at some time subsequent to the time scale of this survey took the game with them into Persia and subsequently round about the beginning of the 16th century CE the Mughal emperors introduced it, or polo, into India. The date or the process by which buzkashi, played with a goat's carcass, became polo, played with a ball, are not known. What is certain is that polo was seen by British residents in the mid-19th century CE in Manipur, on the Indian border with Burma, and the British took it up as a game, forming the first polo club in 1859.

Conclusions

This book has to consider why cavalry took so long to develop after chariotry. Littauer and Crouwel (1979) suggest it is because horses did not wear shoes in the Bronze Age to stop their hooves from splitting as they would with hard work under a cavalryman over rough ground. But it may have just been because driving a horse is easier than fighting from the back of one.

It is worthwhile summarizing here briefly two aspects of chariots in the Bronze Age that have made me treat them by countries. The different types of chariots are too complicated and there are too many types whose design was derived from foreign or several sources or are not known in detail for them to be divided primarily into types. And the dating is too inexact for a chronological division to be feasible.

What is presented here is a drawing together of the accounts and views expressed in the several chapters, in a review of the origins of the chariot, its development into different types that offered contrasting possibilities for use. I present my views, as a military historian who specializes in equestrian topics, on the political reasons for the use of the chariot and its retention beyond its heyday and its replacement by cavalry.

The fast two-wheeled chariot depended for its existence on the domestication of the horse and its taming for man's use. That started in Central Asia, in the steppe grassland that was the animal's natural habitat. There it was used as a source of food, shelter, and transportation by the nomadic tribes who in the 3rd millennium BC eked out a precarious living there. We do not know for certain whether they rode the animal. But the remains of wheels with spokes, essential components of fast horse horse-drawn vehicles, were discovered there. Although a horse could be ridden by the 3rd millennium BC and was directed by a bit in its mouth, not unlike the ones used nowadays, yet without a saddle built on a solid wooden framework a rider's seat would not be secure enough for a man to fight on horseback, nor for disciplined cavalry to be formed. With the primitive technology of the time it was possible only to use a horse to pull a wheeled vehicle. Whether this was fast and light enough to be used as a fighting vehicle, in effect a chariot, is not known. But it would have been possible.

The next evidence, in chronological order, comes from much further south, from Sumer in Mesopotamia from about 2900 BC. The people of the city states there were literate and have left us written and archaeological evidence that they had military wheeled vehicles that we today call battle wagons. Since the horse did not exist as far south as Sumer at that time, they had their heavy fighting vehicles drawn by the local equid, the half-ass or onager.

Then there is a gap of several centuries in the archaeological evidence, until in the 2nd millennium BC, in the Bronze Age, the light fast two-wheeled fighting vehicle drawn by two horses appeared in Syria and Mesopotamia. We do not know who first invented it, although at present the evidence points to Syria as its area of origin.

It spread rapidly in an area that was in constant turmoil with continual wars between the great powers of Egypt, Mitanni, the Hittites and the Assyrians, with the help of their vassals and allies. They fought to control the trade routes that criss-crossed the Levant in order to preserve their wealth and acquire more. New immigrants into the region such as the Israelites and the Philistines fought wars to secure for themselves land in which to settle. At that time of technological change warfare went from being slow-moving infantry battles into fast encounters between armies of many hundreds of chariots.

Every state whose rulers aspired to make it a world power acquired hundreds of expensive chariots. Lesser states that were caught in the middle between the conflicts of the Great Powers followed suit for self-protection.

We are severely limited in our knowledge of how the new technology of the chariot was used strategically and tactically. What we have are the bombastic political records of victories, with the number of chariots involved being given prominent mention. We have to fill in the lacunae in the accounts of battles from our knowledge of evidence from later more literate times as to how military encounters were actually fought.

The conclusion that one must come to is that chariots were prestigious, wildly expensive, and much cherished, although their actual military value was uncertain. But a state had to have them, even if the benefit was political rather than military.

The knowledge of the technology spread beyond the Levant and the Middle East to eastern Europe and China. We do not know with any degree of certainty how knowledge of the chariot entered Mycenaean Greece, but it was probably from Syria.

This was the 2nd millennium BC, several centuries before Classical Greece and even before Homer, who wrote about a period that was a few centuries before his own time.

Excavation in the hills of Armenia, in the eastern part of southern central Asia, and in the southern part of northern China has produced the remains of chariots that are remarkable for their similarity to each other. Yet there is a huge difference between those vehicles and the chariots that are known of from the Middle East, the Aegean and Asia Minor. The best estimate has to be that the idea of the chariot was somehow carried from Armenia thousands of kilometres to China.

Current evidence is that cavalry, that is men fighting together on horseback as a disciplined military unit, first appeared in the Assyrian Empire in the 9th century BC in the form of mounted horse archers.

By the time of the Assyrian defeat of the Elamites at the Ulai River in 653 BC in the reign of Ashurbanipal, Assyrian cavalry had progressed to include mounted men holding spears. It is remarkable that the riders sat only on folded fleeces. These would not have given them the stability necessary for them to handle a weapon that a saddle with a solid wooden tree would have provided. But that was to be an invention that is generally attributed to the Celts in the time of the Roman Empire several centuries later.

The Assyrians do not give reasons for their introduction of cavalry but it is noticeable that in the later years of their empire they were fighting in hills through which it was hard to drive wheeled vehicles but which mounted troops could cross with less difficulty. Cavalry was also cheaper than chariotry and each horseman could carry more weapons on one horse than could the crew of a chariot behind two.

A classification of chariots by types gives us some slight clue to how they were used tactically in the face of the enemy by different countries. The principal distinction was between light chariots and heavy ones. The lightest of light chariots were the Egyptian ones. This would suggest that the emphasis was on speed at the expense of protection for the crew and therefore of the employment of harassing swoops on the enemy rather than charges to close combat. Their chariots with their light construction and slight wheels would also have been of use particularly on level plains but the wheels would have suffered greatly on rocky ground. At the same time the heavier Hittite chariots would have been more resistant to damage in rough ground.

Why did the Assyrians continue to employ heavy chariots that could cross the hills only with the greatest difficulty?

But all empires use a lot of bluff. They pretend to be more powerful than they actually are. It deters those who would attack them, makes theirs functionaries feel important, keeps the subject peoples impressed and cowed, and is cheaper than war. The Assyrians certainly did it with their policy of 'deliberate frightfulness'. Perhaps the ever larger chariots were part of the imperial bluff.

Any study of military history has to examine the evidence against the background of the current technology and politics. That takes us back to Clausewitz; war is policy by different means.

Chronological Table

Rulers of Assyria, the Hittites, Mitanni and Egypt and Chinese Dynasties

Mesopotamia

Kings of Assyria

1179–1134	Ashur-dan I
c.1133–1132	Ninurta-Tukulti-Ashur
1133–1116	Ashur-resh-ishi
1115–1077	Tiglath-Pileser I
	Asharid-apal-Ekur II
1074–1057	Ashur-bel-kala
	Shamshi-adad IV
c. 1050–1032	Ashurnaṣirpal I
1031–1022	Shalmaneser II
	Ashur-nirari IV
1016–973	Ashur-rabi II
	Ashur-resh-ishi II
966–935	Tiglath-pileser II
934–912	Ashur-dan II
911–891	Adad-nirari II
890–884	Tukulti Ninurta II
883–859	Ashurnasirpal II
858–824	Shalmaneser III
823–811	Shamsi-Adad V
810–783	Adad-nirari III
782–772	Shalmaneser IV
753–746	Ashur-nirari V
744–727	Tiglath-pileser III

726–722	Shalmaneser V
721–705	Sargon II
704–681	Sennacherib – The Siege of Jerusalem 701 BC
680–669	Esarhaddon
668–626	Ashurbanipal
625–623?	Ashur-etillu-ili
622?–612	Sin-shar-ishkun
612–609	*Assyria conquered by the Medes and the Babylonians*

Babylonia (the Neo-Babylonian Period)

721–710	Merodach-Baladan II
	3 kings (703–700)
699–694	Ashur-nadin-Shumi
	2 kings (693–689)
668–648	Shamash-shuma-ukin
647–627	Kandalanu
625–605	Nabû-apla-uṣur (Nabopolasser)
604–562	Nebuchadrezzar II (Nebuchadnezzar)
	Evil-Merodach
	Neriglissar
c. 350	Nabû-na'id (Nabonidus)

539 Babylon conquered by Cyrus

Anatolia

Kings of the Hittites

1286–1265	Hattusilis III
1265–1235	Tudhalyas IV
1235–1215	Arnuwandas III
1215–?	Suppiluliumas II

Urartu

c. 850	Arame
832–825	Sardur I
824–806	Ishpuini
805–788	Menua

787–766	Argishti I
765–733	Sardur II
730–714	Rusa I
714–?	Argishti II
	Rusa II
	Sardur III
	Rusa III

Urartu conquered by the Medes

Northern Syria

Mitanni

Contemporary with the XVIIIth Dynasty

c. 1530	Parattarna
c. 1500	Saustatar
c. 1430	Artatama
c. 1400	Shuttarna II
	Tushratta
	Mattiwaza

Contemporary with the XIXth Dynasty

	Shattuara I
	Wasasatta
	Shattuara II

Egypt

Pharaohs of Egypt in the New Kingdom

Dynasty XVIII

1576–1546	Amosis
1546–1526	Amenophis I
1526–1512	Thutmosis I
1512–1504	Thutmosis II
1504–1450	Thutmosis III
1450–1425	Amenophis II

1425–1417	Thutmosis IV
1417–1379	Amenophis III
1379–1362	Amenophis IV (Akhenaten)
1361–1352	Tut-ankh-Amon
1352–1348	Ay
1348–1320	Horemheb

Dynasty XIX

1320–1318	Ramesses I – Battle of Qadesh 1300 BC
1318–1304	Sethos I
1304–1237	Ramesses II The Exodus?
1237–1209	Merneptah

Dynasty XX

1198–1166	Ramesses III
1166–1085	Ramesses IV–XI

Palestine

United Kingdom of Israel

1030–1010	Saul
1010–970	David
970–931	Solomon

Kingdom of Judah (Capital Jerusalem)

931–913	Rehoboham
911–870	Abijam
	Asa
870–848	Jeoshaphat
848–841	Joram
740–736	Jotham
736–716	Achaz
716–687	Ezechiah
687–641	Manasseh
	Amon
640–609	Josiah
	Jeoahaz

	Jeoiakim
	Jeoiakin
598–587	Zedekiah

587 Jerusalem conquered by Nebuchadrezzar

Kingdom of Israel (Capital Samaria).

931–910	Jereboam I
	Nadah
909–886	Baasa
886–885	Ela
	Zimri
885–874	Omri
874–853	Ahab
	Ahaziah
	Joram
841–814	Jehu
814–798	Jehahaz
793–783	Jeooash
783–743	Jeroboam II
743–738	Menahem
	Peka
732–724	Hoshea

722 Samaria conquered by the Assyrians

Chinese dynasties and periods

The two alternative names for dynasties are because there are two principal systems of transliterating Chinese ideograms into the Roman alphabet. The dates for the establishment and demise of dynasties are selected as being the most reasonable ones. The rise and fall of dynasties were not tidy affairs and they often overlap.

Legendary Sage Emperors	2852–2255 BC
Xia or Hsia	2205–1766 BC
Shang	1766–1045 BC
Chou	

Western Zhou or Chou	1045–770 BC
Eastern Chou	770–256 BC
Spring and Autumn period	772–481 BC
Warring States	403–221 BC
Qin or Ch'in	221–207 BC
Former Han (Western Han)	208 BC–8 CE
Later Han (Eastern Han)	23–220 CE

Appendix A

The Wheels of War

This is an account of the experimental archaeology project by the author to design, supervise the building of, and drive, a replica of the Sumerian battle wagon illustrated on the Standard of Ur in order to determine what could have been its performance. The result was filmed by the BBC and shown on television as part of its Chronicle series of archaeological documentary programmes as 'The Wheels of War'.

In 1972 I was lecturing in history at Whitelands College, a teacher training college in west London that is now part of Roehampton University. I was already a qualified archaeologist. But apart from employment as a local archaeologist in government service you don't actually get paid for doing archaeological research or excavation. You have to have a job as well. Usually it's either working in a museum or teaching.

At the same time I was studying part-time for a PhD degree at the Institute of Archaeology of the University of London under Professor Seton Lloyd. The title of my thesis was 'The Development of Transport in Ancient Mesopotamia'. The greater part of this dealt with land transportation from the earliest evidence for wheeled vehicles in Mesopotamia from the Uruk period in the 3rd millennium BC in Mesopotamia to the end of the Assyrian Empire in 612 BC. That covered the Sumerian Early Dynastic period and then went on to the Assyrian Empire with its large army of heavy chariots and its complicated bureaucracy that distributed to various state departments the horses and chariots that the Assyrians had captured in their annual campaigns. So it included the period from the Early Protoliterate culture of Uruk level IV in the first days of the Sumerian city states around 3000 BC to the fall of the Assyrian Empire in 612 BC. A single chapter was on transport by water in that long historical period.

A considerable part of my research centred on the design and use of the Sumerian four-wheeled battle wagons drawn by onagers in the comparatively short time between the later part of the late Early Dynastic II period and the subsequent Early Dynastic III period. Usually for convenience they are abbreviated by archaeologists to ED II and ED III. Both are in the first half of the 3rd millennium BC. These battle wagons are well illustrated on the inlaid sound box known to archaeology as the Standard of Ur that was discovered at Ur by Sir Leonard Woolley in the 1920s CE and are also painted on a vase excavated at the Sumerian site of Khafaje.

I was in the final stages of completing the thesis and as a keen horseman with an interest in experimental archaeology I had a strong notion to build one of these fascinating vehicles and drive it in order to find answers to some of the problems that were at that time still matters of theory and subjects of speculation. We did not know whether the axles were fixed and the wheels revolved on them or whether the wheels were pinned to the axles which revolved under the floor. We wanted to know how easy was it to control the vehicle and its team of four onagers. We had no idea what speed it could reach and how tightly it could turn. In sum we wanted to know how effective it would be as a fighting vehicle.

At that time experimental archaeology, making and using objects that people had in antiquity in order to find out what were the problems and limitations involved with them, was in its infancy. It would cost money and involve a great deal of work. I needed a collaborator. Margo Stout, a research student at the Institute whose field of study was Sumerian bronze work agreed to come in with me on the project. After several possible sources of financial support were rejected I was fortunately able to interest the late Paul Johnstone, Executive Producer of the BBC Archaeology and History Unit and producer of the BBCTV archaeological documentary Chronicle programmes in my battle wagon project. He had a great interest in archaeology, particularly in the experimental side of it and had been involved in several projects of this kind before. He undertook that the BBC would finance the building of the vehicle and my obtaining and training animals to pull it if I could find a woodworker to build it, find a place to drive it, and recruit helpers. In return I would be available to make a series of drives with the vehicle that would form part of a television documentary programme on the Sumerian battle wagon for the Chronicle series.

Finding a stretch of level grass where I could drive the vehicle, there being no desert in west Essex, was not easy. But Mr S. Dixon, Headmaster of St John's School, Epping, offered the use of the school playing fields for me to keep the four animals that would pull the battle wagon, and the services of Dennis Haynes, Head of the Craft Department, to oversee the building of it. Geoff Doughty, the imaginative and unflappable woodwork master, would actually build it with the assistance of one of his fifth form classes. Pupils at St John's, who included my son Rick and my daughter Alison, would provide the nucleus of the considerable ground crew that was needed to feed and water the animals, do the running repairs that were necessary to keep an animal powered ancient fighting vehicle in running order, and in Rick's case act as crew behind me as the driver.

The vehicle would be made of Iroko (*Chlorophoro excelsa*) a West African hardwood of a density comparable to that of the hardwood such as cedar or acacia that the builders of the original wagon could have used. The axles were to be made of beech and the draught pole of a holly tree that had a natural bend similar to that of the Sumerian pole shown on the Standard. The timbers would be fastened together with mortise and tenon joints additionally reinforced with dowels, all techniques known in antiquity. The sides of the Sumerian original were panelled with either basket work or leather. Heavy vinyl upholstery material was used on the replica.

No attempt was made to use ancient Sumerian tools or carpentry methods. Those were not what we were testing. They would have made the job last for years as we would have had to research ancient tools, make them and learn to use them, and retrain Geoff and his pupils as Middle Bronze Age carriage builders.

First of all I had to find animals to pull the battle wagon. The Mesopotamian Onager (*Equus hemionus hemippis*) that the Sumerians used has been extinct since 1917 CE thanks to British soldiers' fondness in Mesopotamia in World War I for taking a pot shot at convenient targets, so the Mesopotamian onager was now no more. The few Persian onagers in London Zoo were not available and anyway were said by their keepers to be untameable. So I had to use the nearest available alternative, the donkey (*Equus asinus*), a relation of the onagers that is arguably of Sudanese origin. I hoped fervently that four of them would be strong enough to pull the vehicle while being lively enough to replicate Sumerian wild asses. So I bought four young donkeys aged between four and six years from a horse dealer. They varied from reasonably tame by donkey standards but unschooled to highly nervous and completely unbroken.

I now had Alfie, Bruno, Dougal, and Cinnamon, two jacks (entire males) and two geldings. Their names were given them by the girls who were students at St John's and had generously offered to feed and water the animals in the school sports field. With hindsight, that great teacher, I had made my first mistake. The ancient peoples of Mesopotamia always had one animal more then was needed for a team in case of accidents. But I had not anticipated that one might be of such a highly nervous disposition that it was inclined to bolt at the slightest incident. And anyway I did not have the money to buy more than the exact number of animals I needed. Alfie was the wild one. And with herd animals, if one bolts, they all bolt. But these were supposed to be wild asses, not tame donkeys. And within a month the wilder donkeys, Alfie and Bruno, would come to hand when called. If slow, that was progress.

There was always the lurking fear that the animals might not be strong enough to pull a heavy battle wagon whose exact weight was as yet unknown. Paul Johnstone and I had deep discussions on what I should do if the donkeys strained and heaved and the battle wagon stayed where it was at the start of the test track. I had to concoct a Plan B. But in the event I need not have worried on that count.

Although I was an experienced rider of horses I had never broken an animal to drive and had never dealt with donkeys. So with the only available book on the subject, the 1937 edition of the *British War Office Manual of Horsemastership, Equitation and Animal Transport* ready to hand I set out to train the donkeys to play the parts of onagers pulling a battle wagon.

Rogue horses end up as dog food before someone gets killed. Horses are bred to do what they are told. Donkeys are bred to produce donkeys. I soon realized that Alfie and Bruno had possibly never seen a human being at close quarters before they came into my keeping, but I had just to soldier on. I certainly had some wild asses. Maybe I was now experiencing the tribulations that the Sumerians four thousand years before had suffered.

It took all my determination, patience and love for animals to keep going with my highly intelligent and strong-minded donkeys. Every day my thirteen-year-old girl helpers shared with me the daily drudgery of carrying armfuls of hay and buckets of water from the school buildings out onto the sports field. Then each day after school we set to with training the donkeys There were a few pieces of asinine behaviour that horse minders in the British army never encountered. Alfie and Bruno had possibly never encountered a human being at close quarters before. A well-known method of approaching a very frightened wild animal is to lie down and crawl towards it on your stomach, for it is only when you stand up that it recognizes you as human and leaps up to flee. Approaching a highly nervous recumbent animal that has its hooves in front of your face is quite a nerve wracking experience. But at some point I had to stand up, for I could not continue with this project in a prone position. So I was careful to get up to at least a crouching position as I reached the danger zone. But patience won to some extent in the end. I persevered with what were the British army's standard methods of training an equid to draught and which are now the generally accepted ones among horse trainers. Mr Cuthbert, the highly skilled saddler in Ray Batchelor's tack shop in Epping made some replica Sumerian head stalls and collars to my design. Here I made my first important archaeological discovery. Ever since 1931 when Commandant Lefebvre des Noëttes wrote his groundbreaking study of ancient horse transport it had been generally accepted that the collars the Sumerians used that did not sit on the bony shoulders like our horse collars half-strangled the onagers. I now knew that was not so. An equid's wind pipe is buried deep beneath protective muscles. When coloured streamers of the kind that appear in the Standard of Ur were attached to the collars they made a very attractive display and looked very Sumerian. To my surprise and relief they did not increase the donkeys' excitement as I had feared it might. Or at least it did not make it any worse.

Working from my full scale drawings of the head collars that the onagers on the Standard of Ur were wearing, Mr Cuthbert made four copies in flexible leather. A problem that had to be resolved was the way in which the reins should be attached to the animals' heads. The Standard of Ur and other Sumerian representations of draught equids show the single rein from each animal attached to a ring that is threaded through either the animal's upper lip or the cartilage of the septum of its nose. I feel strongly that it would have been through the nose as a ring through the lip would have easily torn away.

Clearly a Sumerian type ring was unacceptable on grounds of its cruelty. Another possibility was to fit the donkeys with bits in their mouths like horses. But expense apart, and I had to watch the finances carefully, I did not relish doing this with mature excitable animals that were half-wild. Even if I could succeed in doing it, which was not at all certain, a bit only tells an animal which way you want to turn. It does not force it to obey you if it doesn't want to. So what I did was to get Mr Cuthbert to sew D rings onto the sides of the head stalls and buckle the long webbing reins, one to each animal, on to these. The real direction would be provided by my helpers who would run with the battle wagon and hang onto other reins that were attached to the outside ends of the long yoke that went over the shoulders of the four donkeys. I hoped that at only thirteen years or so of age the ground crew would be strong enough to provide some guidance for the animals. This, as will be shown, they did remarkably well.

It is easy enough to get people to turn up to take part in the film when a camera crew is there, but these girls, prominent among whom was Alison, turned up every day for long weeks for the messy and unromantic business of trudging out across the sports field with buckets of water.

Tethering the donkeys out in the field under supervision by day was easy enough. The problem of what to do with them at night was more complicated, since we had no stables. I started in approved cavalry style by laying out a horse line of a rope between two stakes and attaching to it at intervals the leading ropes from the modern head collars the donkeys wore when they weren't pretending to be onagers. This must work with cavalry horses. It doesn't work with donkeys. They soon got their ropes in a terrible tangle and made the night hideous with their braying. This produced great protests at my cruelty to animals from the inhabitants of the houses surrounding the school. But their concern for the donkeys' happiness did not extend to their coming out in a dark winter night and helping me and my immediate family sort out the cats' cradle of ropes that the donkeys had got themselves into.

Next day I bought an electric fence that ran on a dry battery and it worked beautifully, with the donkeys in a nice paddock that could be moved to fresh grass every couple of days. I distinctly remember one incident that brought home to me just how smart donkeys are. If you touched the fence, you got a stinging shock and the donkeys kept at least three feet away from it. But they would still like if possible to get out of the paddock to where the grass was greener on the other side. Dougal, my largest and most beautiful multi-coloured jack, was the nearest I could get to being a pet and he was also very bright. He wanted out of the paddock but he didn't want an electric shock. Someone else could get that. So one afternoon he took off in a gallop round and round the paddock six or eight feet back from the fence. Another of the donkeys joined in, galloping round between Dougal and the fence. If an equid breaks into a gallop, another nearby one joins in. He doesn't enquire where the danger is. When Dougal had his work mate nicely positioned between him and the fence he charged and shouldered him it into the fence. His mate got the shock, while knocking the fence flat on the grass. Dougal then jumped the fallen fence and was off for pastures new.

I had a two-wheeled training vehicle made from of slim metal rods and bicycle wheels and with care could drive all the donkeys in single harness round the sports field. I was beginning to feel I was getting somewhere. Until I took Alfie, my nervous donkey, for a drive. He was as good as gold until I reined in to a halt, ready to dismount. Then a pigeon took off in a clatter of wings twenty yards in front of us and Alfie knocked down the two handlers holding his head and was off. Like every good rider I knew all the tricks for stopping a runaway ridden horse, but they don't apply when you are driving.

All you can do is sit tight and hope for the best. The light metal practice cart soon turned over. The shafts sheared off as they were designed to do in an emergency and the body of the cart I was sitting on was dragged along the grass at speed on its side behind an excited donkey, breaking up as we went. The long reins were separate and not fastened together in a loop and the spare ends of them were coiled up underneath me on the seat out of the way. I remember them snaking out from beneath me at high speed in a most alarming fashion as I hoped fervently I didn't get castrated. The cart came to a halt and with the reins now no longer in my hands, Alfie bolted for the other side of the sports field until, without the cart behind him, he stopped.

With the light training cart now a twisted wreck we carried on without it. The main part of my donkey training now was long reining them where

I walked behind the donkeys holding the six-metre long reins as we walked round the sports field. After six weeks the tamest of the donkeys, Dougal and Cinnamon, would canter on command and would walk side by side under a yoke. They were not a problem. Alfie was. I should just have to work extra hard on him.

I had lost quite a lot of skin in the spill but was otherwise unhurt and Alfie seemed unperturbed by his experience . While I was involved in these exciting incidents with the donkeys, Geoff Doughty was building the battle wagon with a fifth form class to my design.

Working from the inlaid pictures of the battle wagons in the Standard of Ur I had three factors to bear in mind in deciding what size to make the replica. There was the height of an onager, one metre (10 hands high) at the withers, the diameters of excavated Sumerian wheels, which ranged from 0.50 to 1.05 metres, and the average height of a man, which I took at 5 feet 7 inches (1.72 m). By combining these, a size for our battle wagon was calculated that proved satisfactory. I had worked on the general assumption among archaeologists that in the Standard of Ur the front of the battle wagon was portrayed as if it was part of the side. So I could estimate the width of the vehicle. While this was going on the film director, Antonia Benedek, and camera crews came regularly to St John's to film the progress of the donkey training and the building of the vehicle.

When it was completed the replica certainly looked an almost exact copy of the vehicles on the Standard of Ur. I felt I had got it right there.

The overall length of the whole equipage from the tip of the draught pole to the back end of the body was 16.4 feet (5.00 m) and the width of the team of four animals was 8.2 feet (2.50 m). The body of the wagon was made in two halves, the floor being two planks fastened together with dowels. The reason for this was that planks of 1.97 feet (0.60 m) in width, the width of the floor, were not available in Iroko. The sides and front were built as three separate units and they were fastened together and to the floor with mortise and tenon joints. The suggestion has been later been made to me that the floor should have been made, like that of Egyptian chariots, of interlaced leather straps to act as a rudimentary sort of spring. But there is no evidence for that and the Egyptian chariots were five hundred years later and fifteen hundred miles away across the desert in a completely different country. The empty battle wagon without animals and human crew weighed 342 pounds (155 kilograms). We weighed it by loading it on a lorry and driving it onto a weighbridge, which weighed only lorries, and weighing them together. Then we offloaded the battle wagon and weighed the lorry without it.

On studying the Standard of Ur in greater detail it became clear to us that there was a good reason why the wheels were not made of three planks side by side. We knew that modern solid wheels on farm cards in the Middle East were made from three planks fastened side by side and surviving soil impressions of ancient solid wheels of Sumerian times from the city of Susa south-east of Sumer are of three planks. It was pointed out by Dennis Haynes that if the wheels had been made of three planks the centre one would have extended diametrically to the circumference, so twice in every revolution the weight of the vehicle would have come on the grain of the wood. The lozenge-shaped central part of the wheel reduced this stress by taking some of the weight and distributing the pressure by its shape. Another reason for its presence was because the heart wood in the centre of a tree was too soft to take the pressure of the axle, so it was replaced by a piece of harder wood with a straight grain. We were learning about the problems the Sumerians met and solved them as we went on.

For the first tests the axle was lashed to the underside of the floor with leather thongs and the wheels revolved on the well greased axles. The type of axle was a vital part of the research I was intent on undertaking. Driving a battle wagon at speed was incidental. I knew from looking at primitive farm cars and wagons in the Middle East that there were two ways in which the axles and wheels could be organized. The axle could be fixed rigidly to the vehicle and the wheels revolved on the ends of it. Or the wheels could be pinned immovably to the axle, which revolved under the vehicle. We did not know which configuration the Sumerians used. A war vehicle like the battle wagon would have to be able to execute tight turns to get back out of enemy range and if one of the possible configurations made a much tighter turn possible, then it was probable that it was the one that the Sumerians used. So I would drive the battle wagon into a tight turn with first a fixed axle and then again with a revolving one.

It was discovered that the hubs of the wheels seen on the Standard, which must have fitted over the axles and inside the centres of the wheels, did nothing to strengthen the wheels or hold them together. They were left as freely rotating bearings inside the wheels, which reduced the friction and the strain on the lynch pins that passed in the usual fashion through slots in the axles and kept the wheels on.

The pole, the link with the double bend between the wagon and the yoke, was not easy to find. It had to have just the right S-bend to fit under the wagon and carry the yoke over the donkeys' shoulders. Such a pole, of holly, was found after a lengthy search in a forest. But how the Sumerians assured themselves of an adequate supply of suitably shaped poles is not known. The pole was fastened to the underside of the vehicle by passing it through a wooden collar pegged to the underside of the vehicle. A transverse dowel through the pole behind the collar took the pull. Ideally the collar should have been pegged to the underside of the floor, but in fact in the first fixed axle version it was fastened to the underside of the front axle because the pole did not have quite the right curvature to carry the yoke over the animals' shoulders. Some concession to the practicalities always has to be made in a project like this. If it was a question of my neck or strict historical accuracy then historical accuracy came second. For by now I was under no illusions about how dangerous was the enterprise I was undertaking. I wonder what modifications from the ideal the Sumerians had been forced to make.

Ahead of the collar the pole was lashed to the floor to stop it from swinging sideways. The Standard shows no girths round the animals' middles, but they had to be used. It was found impossible to fix the pole so rigidly that the yoke was held off the animals' shoulders. It sagged onto them and when the donkeys cantered, the considerable up and down movement banged the donkeys' shoulders and alarmed them enough to make driving impossible. So the yoke was padded with foam plastic and strapped firmly down via the girths onto the animals' shoulders. This solved our problem very well. It was suggested to me by various colleagues after the project was ended that in a four-wheeled vehicle that does not rock with the animals the pole should have been hinged vertically so that it would have gone up and down with the animals as they cantered. I agree with them. But I cannot see how that would have helped much without the girths to which the yoke could be lashed. We had the prototype. This had not been done before. But then we were doing academic research and making a television documentary. The

Sumerian young men were fighting a war in which fatalities were inevitable and accepted. Wars are fought by eighteen-year-olds who think it will never happen to them.

Part way through the tests the axles were modified so that they revolved in axle boxes and the wheels were pinned to the axles so that they revolved with them. There had been much speculation among specialists in the subject as to which type of axle the Sumerians used, so both were tested to see what was the difference in friction and the vehicle's turning circle.

In the turning axle version the pole was fastened to the wagon by means of two wooden collars pegged to the underside of the floor. Dowels through the pole stopped it from revolving about its own axis. When high speed turns were done with this version, it was found that the dowels holding the two halves of the forward collar together were repeatedly broken. For speedy practical convenience they were replaced with steel bolts. This recalls the two so-called bolts that Woolley found in the earth associated with a wagon at Ur. It is possible that the builders of the original wagons had the same problem and solved it in the same way.

We wished to test the efficiency of the Sumerian type of draught collar, as Woolley had expressed his belief that such collars would press on the jugular vein and deprive the animals of half their strength. Our veterinary surgeon disagreed, saying that the large blood vessels were deep, and well protected by muscles, and he thought that the collars would not inhibit breathing. It was found out by experimentation that this was the case. The harder the animals pulled, the more the collars sat down on their shoulders, moving away from the neck completely. Each donkey of the team had to maintain a pull of just 11 to 17 pounds (5–8 kilograms) at any speed.

While the soft leather replica Sumerian head collars that Mr Cuthbert made for us were efficient as muzzles, keeping the donkeys mouth closed, they in no way restricted the nasal passages. That was important as equids breathe solely through their noses, never their mouths. But the replica Sumerian head collars were too soft to serve as a medium of guidance. So for safety in the trial runs, bits being impossible and nose rings unacceptable, the donkeys were run in horse training cavessons of the kind used for lunging horses. These are a type of bridle with a heavily padded steel strip over the bridge of the nose and several rings to which reins can be buckled.

We tried various combinations of reins, two or four, but not the Sumerian version with two reins passed through rings on the central draught pole. That clearly would not work with the cavessons. While wishing to retain as much historical accuracy as possible, I had perhaps a greater determination than had the young Sumerian soldiers to survive the project.

It soon became obvious that the donkeys would bolt whatever I did and there was no way of stopping them. No man is strong enough to stop four donkeys that do not want to stop. The standard method of stopping a runaway horse is to run it in circles until it gives up and comes to a halt, which it will in the end, after several circuits. This was not feasible in a wheeled vehicle that would turn over.

Our method with the cavessons still gave little control so for safety we would have grooms on foot running beside the outside donkeys and holding additional reins with which they might exert some pressure on the animals. But that could not be guaranteed to bring the battle wagon to a halt. I mulled over and rejected several ideas. What I needed was some way of separating the animals from the vehicle in an emergency. With a two-wheeled chariot that would lead to a smash but a four-wheeled battle wagon should, if all went well, stay upright on its wheels. Dennis Haynes came up with the solution. This was a modification of the friction free barrel latch that is used to drop bombs from aircraft and depth charges from the sterns of destroyers. He built one. The catch, a complicated and beautiful piece of engineering about six inches square was bolted on top of the end of the draught pole. From it two short lengths of motor cycle chain were looped up over the yoke and back to the catch, holding the yoke firmly attached to the draught pole. Inside the front of the wagon was a steel grab handle that was connected by means of a wire to the barrel latch. In an emergency, which turned out to be every run, I threw the reins overboard and pulled the handle. The ends of the motor cycle chains flew up in the air, a reassuring sight, as they were released from their fixture in the latch and the yoke was now separated from the wagon. The battle wagon trundled to a halt, if we happened to be travelling in a straight line, and the donkeys cantered off and quickly lost interest in rushing about when there was all that grass around for them to eat. There was a small point to be considered. The grab handle had to be pulled slowly to give the latch time to work. If you pulled it too quickly or if the tension on the wire connecting it to the latch was not right, the latch out on the draught pole could revolve past top dead centre and lock up. I never got into the wagon without first testing the dropper three times to make sure that it worked. Only once the tension on the wire was not right. That would have been curtains for me if I had not seen the necessity to adjust the length of the wire from the handle to the latch.

The test track was a hundred and fifty yards long stretch of sports field slightly up hill with an adequate run-out area at the end. To give me something to aim at and help me judge how straight I was driving I laid out a trail of sawdust up the centre of the track and planted a pole beside the track at every fifty yards.

The battle wagon had its first slow runs under the human power of my young helpers and without the donkeys to ensure that it would go in a straight line and for me to see what would happen when it was pulled round in a turn. That went, as anticipated, without incident. Now we would just have to see what happened behind the donkeys, who would be going much faster.

So on 9 September 1975 with the camera rolling and Antonia and Magnus Magnusson, who would be the presenter and commentator in the film, standing by, I pulled on the yellow motor cyclist's helmet I had purchased and climbed aboard the battle wagon for test number 1, feeling rather as if I was Donald Campbell about to attempt the world speed record.

Magnus Magnusson was a delightful man. He had said to me, 'My job is to present you in the best possible light.' That cheered me up tremendously. No matter what shaming disasters might occur, I should get support from one quarter.

Behind me as crew was my fourteen-year-old son Rick, chosen as he was a capable rider and not given to panicking. I shook the reins, clicked my tongue, and told the donkeys to walk on. They did that and after a few yards the combined efforts of the reins and my helpers hanging on to the straps that were tied to the ends of the yoke pulled them to a standstill. At an all-up weight of 304 kilograms we had done 45.6 yards at a speed of 2.45 miles an hour. That was good.

Test number 2 was more exciting. Without difficulty the donkeys were encouraged to hit a canter.

We now knew what we had set out to discover, the battle wagon's speed and its turning circle with the fixed axle configuration. The fact that it was so bumpy that if the driver steadied himself with his arms on the front of the vehicle he saw double, one above the other, could be endured. It only remained for us to discover what difference the rotating axle made in the later tests and how efficient the vehicle was as a platform from which the warrior behind the driver could throw javelins. We knew what the battle wagon on its own was capable of. What we did not know, and could not test, was what happened when a mass of battle wagons operated at speed together. For a start we did not know how many battle wagons a Sumerian city had and how they were employed; whether they attacked in line ahead, one after the other, or in line abreast like a cavalry charge. Short of the extremely unlikely fortuitous discovery of a Sumerian cuneiform text providing that information we would never know. And we did not have that kind of information from the much better written up Assyrian Empire. So this would remain unknown for the probable future.

Also we did not know what spills, smashes, collisions and casualties the Sumerian battle wagon crews had, in training or in action. It is generally accepted that with western European armies they give up completely when the casualties reach 50 per cent. But the Japanese in World War II kept on fighting with a worse casualty rate. German U-Boat crews in the same conflict suffered a 40 per cent casualty rate and kept on. British RAF bomber crews had a 5 per cent casualty rate on each raid so that statistically they were dead after twenty trips of a thirty-raid tour but few cracked. In the First World War the life expectancy of a British infantry second lieutenant was three weeks. But men still volunteered for promotion to that rank. A factor in maintained morale is now known to be whether men can see what was happening to others. If they could not see their disasters they were insulated from the reality of the situation. As a former infantry officer who had seen action I knew the mixed jumble of emotions of fear and excitement and the heavy load of responsibility for others that stop you thinking about yourself. But we could not tell the situation that Sumerian battle wagon crews faced. And so we carried on.

The third run was memorable. Predictably, my excitable donkey bolted and the others followed suit one after the other along the yoke. The battle wagon went into a right-hand turn. It was time to drop the animals. Now. Immediately.

To get a good grip on the reins I had them held tightly in my fingers and now I was pulled forward against the inside of the front of the vehicle and my hands were trapped so that I could not let go the reins and reach the release handle. I grunted to Rick that I could not release the animals. Without a qualm he said, 'Don't worry, Dad,' reached over my shoulder and pulled the release handle. The donkeys relieved of the presence of this nasty rumbling thing that was pursuing them trotted happily off and the battle wagon trundled to a halt. Rick was the kind of crew you needed. Reliable in an emergency.

The next run a day later was even more exciting After the excitement of the previous run, Rick and I intended to go only at a walk but continue right up the test track to the end. If the donkeys took off I should not immediately disconnect them, but carry on until things got far too dangerous. For if I kept aborting the run as soon as the animals took off I should never find out what I wanted to know; what the tests were all about. But we would only go for a quiet walk to let the donkeys stay calm. Yet I was well aware that I did not know what would really happen. I found out. Within yards. The donkeys could hit a gallop from a halt within a yard, and with that distance covered they were off. As Alfie was on the left outside position, since they were positioned along the yoke by size, with the largest in the middle, runaway turns tended to be to the right. Fortunately a right-hand turn was very slightly up hill and my helpers on the yokes and I clung on in our positions. The battle wagon went into a tight and very fast right-hand turn. I was certainly now finding out what was the tightest turn the battle wagon could do.

I was not going to drop the animals this time, not until I had discovered how tight a hundred and eighty degree half circle the donkeys could drag the battle wagon round. We whizzed round at what I later calculated was twelve miles an hour. That does not sound very fast, but it is when you have no brakes and the engines are stuck at full ahead. By the time we had described three quarters of a circle we were going slightly down hill and ahead of us was a copse of trees with beyond them a five feet drop into a lake. It was time to stop. Crouching on the floor and holding on with my right hand to a strut that connected the front of the wagon to a side rail I pulled the disconnecting handle. In a wheeled vehicle the safest policy is to stay with the vehicle if you turn over. Try to jump out and you are very liable to be crushed under a wheel. Once separated from the wagon the donkeys soon came to a halt. In the wagon we were going round in such a tight circle that centrifugal force took over and we rose into the air and toppled over on our left side. I felt something sliding up my back. That was Rick jumping or being catapulted out of the wagon. Looking at the film of the incident afterwards I saw with interest that we had actually been airborne, eighteen inches off the ground when we turned over. The strut I was holding tore away from its moorings and I was thrown violently across the wagon, hitting the side rail on the left side. Because of an unforeseen accident in that construction, that particular strut had been fastened down only with screws, not a dowel. But it might have torn loose anyway.

We were stationary now and I scrambled out onto the grass. Rick was unhurt and I had suffered no apparent immediate injuries. I was later told by my doctor when the shock wore off and the pain came that I had broken a rib on my left side in two places. But broken ribs mend quickly, although they are very painful when you laugh. More sobering was to see that in being flung across the wagon I had bent the three-eighths of an inch thick steel handle of the disconnector with my head. I had never felt a thing. I was glad that I had strongly resisted the suggestion before the run that I did not wear the rather conspicuous helmet as it did not look so good on camera and we were only going to do a walk anyway. But although you may walk willingly into potentially lethal occupations, remember that your life is in your own hands.

Three more timed walks were done with two animals. Alfie was just too skittish to be used again, so the tests were done thereafter with the two most dependable animals, Dougal and Cinnamon. The times for tests were similar to the times obtained in tests with four animals. This tended to confirm what has been my theory that although the augmentation of a chariot team from two to four animals might increase the distance over which animals could pull the vehicle, its range of operations, it has no effect on the speed at which they can pull it.

By connecting a spring balance between the animals and the vehicle we found the force in kilograms necessary to start the vehicle from a halt and to maintain it in motion. That this might be too great for the animals had been my fear in the early planning stage of the project, before I discovered how lively and strong the donkeys were. With the fixed axle the pull necessary for the whole team to start the empty battle wagon on dry grass varied from 46 to 55 kilograms (101–121 lb). Thereafter a pull of only 25 to 35 kilograms (55–77 lb) was required to keep the battle wagon in motion. This should not change at faster speeds. To find the pull that each donkey would have to exert these would have to be divided by two or four, depending on the size of the team being used and taking into account the weight of the crew. These figures were well within the capacity of any donkey. They gave a coefficient of friction, the drag on the wagon necessary to overcome the retarding effect of the friction of its moving parts, of 0.1. This is the figure one would expect from a modern motor car. So the use of heavy wooden components without modern metal bearings did not make the battle wagon harder to pull than a modern vehicle of the same weight.

With the turning axles, to our surprise, the coefficient of friction was less than it had been with the fixed axles and the pull necessary to start the vehicle from rest was now from 35 to 40 kilograms and the pull necessary to maintain motion was between 20 and 35 kilograms. These were much less than the figures obtained from tests with the fixed axles with wheels that revolved on them. This surprised us greatly. For in theory the more archaic set up with the rotating axles should involve much more weight to rotate and a great deal more friction to overcome.

And in the first test, number 4, with only two animals they gave me a much faster run than I had had with all four animals. And they kept straight up the track with no runaways and no circles. I was getting the figures I wanted. I can only put it down to the absence of their more excitable comrades. Equids have a very intense, you could almost say steamy, emotional life and perhaps Dougal and Cinnamon were relieved to be quit of the company of Bruno and Alfie.

Three more timed walks were made with just two animals. The times for tests 4–6 were similar to those for tests 1 to 3 with four animals.

I tested the two donkeys for the first time at a trot. When I put them into a turn the turning circle was larger at 22.17 metres than I had got with four donkeys. But this had not been a bolt and during the turn Dougal and

Cinnamon shied away from the direction of the turn, which must inevitably have increased its diameter.

In the eighth test I attempted for the first time to run the two donkeys at a trot with the battle wagon hitched up. The time I got over 100 yards was better than I had got when long reining the team of two under the yoke without the wagon.

Tests 9 and 10 were fast timed runs at a canter and they showed that, at least over a short distance, the weight of the wagon was nothing to the donkeys. The best speed up the measured track, which had a 1in 50 rise in it, was 8.64 mph (13.91 km/hr) and the animals were not running flat out. Two turns were made at a canter. One had a diameter of 26.05 metres with grooms holding the ends of the yoke and another had a diameter of 32.45 metres with no one holding the animals' heads. The donkeys were improving rapidly in controllability.

We managed to cover the 150 yard length of the test track at all equid gaits of walk, trot, canter, and gallop without untoward incidents but with great care. Sometimes I had Rick as crew and sometimes I had one of the other girls behind me. I had substantial employee third party injury insurance but as the crew's employer I was the only one who was not insured. But I never had to claim. It is my regret that my daughter Alison never got a chance to ride on the battle wagon now that I had enough experience to make it reasonably safe. But she was so indefatigable in tending for the animals and so dependable as a leading member of the ground crew that she always seemed to be more valuable in one of those capacities.

Now I never bothered trying to rein in the animals at the end of a run. I just disconnected them. We continued with the tests without any more dangerous incidents.

Paul Johnstone came to see how we were getting on. He was a very sick man with heart trouble but expressed himself happy if I could undertake a couple of more runs like the ones he had just seen. I assured him I could. If not routine, the test runs were now achievable. Paul had been a great help to me on this very emotionally charged project and although he died before the film was transmitted on television, I am pleased that he knew that what was to be his last film was going to be a success.

I now knew that the donkeys could pull the battle wagon at 8 mph and turn it to face in the opposite direction after a semi-circular turn of less than I had expected with the fixed axle.

Now I had to find out what difference if any the rotating axle would make to the turning circle. In theory the turning circle should be greater because the wagon would lack the differential effect of the wheels turning independently with those on the outer side of the turn rotating more quickly than those on the inside. But in practice the wagon was being hauled round sideways by brute force. So the increase of the turn diameter with the rotating axle might be less than would be expected. I realized that the grass of a playing field did not replicate the surface of the Mesopotamian desert between city states. But anything rougher would no doubt have resulted in spills that would have brought the tests to a fault.

The next thing was to convert the test battle wagon to a rotating axle. Geoff did this after he and I had performed a little piece to camera in which we discussed how to manage this very complicated measure. We had already discussed this together in great detail for a couple of hours but to film that would have made any television viewer rush to put the kettle on.

Tests with the rotating axle showed to our surprise that there was no significance difference in the turning circles between the two different types of axles. So we did not have anything that might suggest which axle the Sumerians used. The question was still open, as it is to this day.

But the number of repairs that the fifth formers of the ground crew had to make every morning on split and broken components of the undercarriage where the draught pole joined the body before I could drive the wagon suggested to me that one charge a day was as much as the vehicle could withstand.

It has been suggested that Sumerian onagers were not schooled in the sense that the term is used today. While a single battle wagon is hard to control it is possible, from what is known of later cavalry, that it would be much easier to manage wagons that were crammed together in a charge or surrounded by running infantry. Cavalry horses run straight in a charge because there is not the space to run out.

As I drove the battle wagon it became obvious that the only spear that could be used effectively from it was a thrown one, a javelin. As the team of four animals extended over a metre out beyond the wagon on either side, not even the tallest warrior in the wagon could have reached an attacker with a battle axe or even a spear in his hand. The 2.50 metres width of galloping onagers would have crushed or scared off any opposition in the path of the battle wagon. The only time a soldier on the wagon would be glad to have his axe would be if a fearless and agile enemy managed to leap onto the wagon from behind.

As javelins are shown on the Standard being thrown from the wagon, we concentrated on their use and the effectiveness of the wagon as a throwing platform. The BBC asked Dave Travis, seven times javelin champion of Great Britain, to participate in the experiments. As even the good donkeys would be an unknown quantity with javelins flying about, I gave the animals the day off and hitched the battle wagon by way of a tow rope to the back of my car. Also, Dave would need to do several runs and I had no desire to try doing more than one run a day with the donkeys. So my car took their place.

We had prepared an enemy for Dave to attack. The girls and I had drawn, painted, and cut out four soft ceiling board images of Sumerian infantrymen copied from the foot warriors on the Standard of Ur. And very convincing they looked. They were placed upright in a row with half of them on each side of the track half way up it so that they faced the battle wagon as it passed at speed through the gap between the two halves of the enemy force.

It was now 30 October and a thick clammy mist cut visibility down to just over the length of the test track. Present were Antonia, the camera crew, the battle wagon crew, which would be Margo and Dave, and a reduced ground crew. You could still see the far end of the track from the start. But it combined with what we were doing to give a creepy unreal atmosphere to the whole proceeding.

I engaged first gear and we were off up the track at 8 miles an hour. The aluminium javelins rattled together in the quiver made a great tinny noise that would have added to the donkeys' worries if they had been there. But the battle wagon stayed straight behind the car as I drove up the sawdust trail in the centre of the track. The slight concerns I had experienced that the wagon might have a tendency to weave and zig zag up the track proved groundless.

Four more runs were done like that and Dave had no difficulty in plucking one javelin after another out of the quiver and scoring hits on the nearest Sumerian soldiers at a range of ten yards.

Dave had a most relaxed attitude to this new kind of javelin throwing. For as he remarked, he had only ever thrown javelins for distance, never at targets. Margo had turned up from her supervision of bronze battle axe casting and she was given the front position on the wagon and Dave stood behind her in the warrior's position.

The first problem was where to hang the quiver containing the javelins. From the twisted perspective on the Standard it appeared, although it was not certain, that the quiver was hung on the front of the wagon. Margo did not believe this possible, as the thrower would have had to reach over the driver while holding on to the side of the vehicle at the same time. So for the first run with javelins the quiver was hung on the right side of the wagon between the two wheels.

Dave held on to the left side of the wagon and pulled out and threw the javelins with his right hand. The javelins are shown on the Standard and also on the limestone plaque to be point uppermost in the quiver. Dave demonstrated that to extract a point uppermost javelin from the quiver, turn it round and throw it, involved an extra move that wasted time. But if the javelins were placed point down in the quiver, the extraction and throw could be performed in one continuous movement, allowing greater speed. It must be assumed that the artists who made the Standard and carved the plaque showed the javelins point up so that it was obvious what they were.

After the run with the quiver on the side of the vehicle it was obvious that this would not work. The javelins were awkward to reach and the thrower had to crouch in a cramped squatting position in order to hold on to the side railing. It was impossible to hold on to the low side with the left hand and at the same time make the full sweep with the right arm that was necessary to throw the javelin.

We moved the quiver to the edge of the front of the wagon, but on the right hand side, not the left one as seen on the Standard. The artist must have put it on the left for greater clarity. The butt ends of the javelins now projected out beyond the high front of the wagon. The thrower would have to cope with that.

There remained the problem of what the thrower could hang onto in order to prevent himself from falling under a wheel. The only feature in the wagon high enough to allow him to stand up was the front of the wagon, and this was effectively blocked off from him by the driver. The only effective hand grip for the javelin thrower was the driver's left shoulder. Having worked this out, we had another look at the Standard. It is clear that in the second and fourth 'frame' of the battle wagon sequence the left arm of the soldier behind the driver is raised and his hand disappears behind the driver's left shoulder.

The first of the seven tests with the javelins was designed to tell us how quickly a series of javelins could be thrown. Two of the runs were made at 5 mph and the other four at 10 mph by the car's speedometer. It was found that while the wagon reached a point of what might be termed 'maximum vibrations' at 5 mph, this did not worsen with increased speed beyond that. It was not the smoothest of rides but it neither rattled the teeth nor impaired the vision. A speed test at 15 mph was so frightening for those on the wagon that it was not repeated.

The second test done with the javelins was to find out how accurately a javelin could be thrown from the wagon at speed. Dave could throw one every two or three seconds. Three of the life-sized cutouts of Sumerian soldiers were propped up at 120 metres from the start and between seven and eleven metres off the track to the right while the fourth was placed again at the 120 metres point but between five and seven metres to the left of the track.

I was so pleased with the way things were going that I extended the javelin-throwing tests for a couple of more days and had Dougal and Cinnamon brought up from the paddock and harnessed up to the battle wagon. The first run with the donkeys was done at a walk and the second at a trot. Dougal and Cinnamon flinched momentarily as the first shiny aluminium javelin flashed past them, then ignored the following ones. This was surprising as all along the donkeys had been frightened by unexpected things but now flying javelins and streamers flapping on their chests were causing them no concern.

A total of twenty-two javelins were thrown and eleven hit the targets. This 50 per cent accuracy was quite remarkable, considering that modern javelin throwers were trained to throw to ever greater distances, not to hit targets. It speaks highly for Dave's skill that he was able to do what he did from a moving vehicle without any preliminary practice. One must assume that real Sumerian enemies would not have been so obliging as to stand still while javelins were thrown at them. But perhaps they were packed so closely together that they could not move.

On the last day of the tests, a bright sunny one where the autumnal mist had dissipated, Professor Stuart Piggott came out to St John's to record his views on the battle wagon project. He was a very senior figure indeed in the archaeological world and had written extensive and innovative books and articles on early wheeled vehicles. So it was gratifying indeed for me when he gave his opinion that my research and experiments had added to human knowledge and were an extremely worthwhile project.

When I had just one run left to do I had a vacancy for a crew member on board the battle wagon so I offered a ride in it to anyone of the large group of people who were involved in the exercise. I was hardly knocked down with the response of volunteers and then Professor Piggott called out, 'Right, Duncan, I'll come,' and stepped aboard behind me. I concealed my feeling of awesome responsibility. One of the leading figures in my profession had placed his life in my hands. Nothing had better go wrong. Off we went at a gallop. Nothing did go awry and I was mightily relieved when we arrived safely at the end of the track and inordinately proud that Professor Piggott had put his trust in me.

Appendix B

Who Was Who

This appendix contains readily accessible information on peoples, places, and minor states in Europe, the Middle East and Central and East Asia who are mentioned in the book but which no longer exist. The only clues we have to the original places from which the peoples who pushed into and occupied areas of the Middle East in the 2nd millennium BC come from their languages. For the most part we depend on their being written in a script that we can read. Otherwise we are dependent on an analysis of place and personal or deities' names, but these are often of limited usefulness.

Languages are divided by linguists into families. The two principal ones we are concerned with here are the Semitic and the Indo-European families. Today there are three Semitic languages still spoken, Arabic, modern Hebrew, and Aramaic. The principal defunct ones were Assyrian, Babylonian, Hebrew and Canaanite. All the languages of modern Europe belong to the Indo-European family with the exception of Basque, Hungarian, Finnish and Estonian. The Indo-European family is so called because the first known developed one is to be found in India. That is the now extinct language of Sanskrit, which is the earliest known one of the family. Ancient and now extinct Indo-European languages include the languages of the Hittites. Iranian is a branch of the Indo-European family and includes the modern Farsi of Iran and the Dari of Afghanistan. Some languages such as Elamite and Sumerian are not related to any other known language, so we cannot ascribe them to a family. Just to complicate matters, the ordinary people of some countries such as Mitanni spoke a language of one family while the ruling class spoke a language of a different family.

The descendants of the peoples who were prominent in the 2nd millennium BC but no longer exist have been absorbed by later arrivals. They are now Arabs, Turks, or Kurds, although they would probably not agree with that identification of their original ancestors if you asked them. But then, for romantic and political reasons they might.

Today it is quite common in Middle-Eastern society for a dinner party to be carried on in four languages, where everyone speaks two or three of them.

The Arabs

The definition of who is a modern Arab is a complicated one, to be entered upon with delicacy. It is based on speaking a modern Semitic language, Arabic. Arabs speak Arabic and revere their language. The great majority are Muslims, of various sects that hate each other. However, many are not Muslims and if you explore the byways of Arab society you will come across Arabs who do not speak Arabic.

In the 3rd millennium BC through to the end of the period with which we are here concerned a people who were probably the ancestors of many who are now called Arabs were nomads in the western parts of the Syrian desert. The name Arab comes from an old Semitic root meaning to wander. They were a recurring nuisance to the expanding empires of the Middle East, although useful for their ability to survive in harsh desert conditions. Where possible they were persuaded to work as scouts or messengers for the armies of the settled peoples.

Aramaeans

The Aramaeans were the Semitic-speaking inhabitants of the small coastal and hinterland city states in the north of Palestine and the kingdom of Israel, in modern Syria and the Lebanon, in such towns as Alalah, Arpad, Damascus, Sidon and Ugarit.

Individually they were no military match for the Assyrians until gradually they learned the advantages of forming coalitions.

Amorites

These people were composed of wandering Semitic tribes, also known to modern authorities as East Canaanites and West Semites. They were known to the Sumerians in the time of the Third Dynasty of Ur in the late 3rd and early 2nd millennia BC. By that time they had settled down in small city states in Syria.

Bactria

This was the name of an area of ancient central Asia, to the north of Afghanistan, which was celebrated in the 2nd and 1st millennia BC for the quality of the horses coming from there. Chinese horse breeding started with Bactrian horses. The Assyrians prized Nisean horses, which came from Bactria.

Canaan

Canaan was the name given in antiquity to the southern part of Palestine that was occupied by the Semitic Canaanites until it was seized by the Israelites at the end of the Late Bronze Age, about 1200 BC.

Chaldaeans

This is the name given by the Babylonians to tribes of Aramaic-speaking Amorites who in the late 2nd millennium BC came from the Syrian desert and settled in the south of Babylonia, adjacent to the Gulf. There they adopted a version of the closely related Babylonian language that was intermingled with Aramaic.

Cimmerians

This Indo-European speaking group of horse riding nomads who lived in the Caucasus and north of the Sea of Azof had a reputation for ferocity in the 8th century BC. Their history is known from Assyrian texts. They were driven south by the Scythians into Anatolia, where about 714 BC they attacked Urartu and then went on to conquer Phrygia about 696 BC. Subsequently they conquered Lydia in western central Anatolia. They are significant because they drove other nomad people south of them further south. They do not themselves come into this history, but they were the force that caused other nomad peoples south of them to push further into the Assyrian area. Eventually they themselves were conquered by the Scythians.

Edom

This was a Semitic kingdom to the south of Moab, which itself was on the eastern side of the Dead Sea, across from Judah, which was on the western side of the Dead Sea. Edom was large and at its greatest extent it spread west to the south of Judah and southwards to the Gulf. It was mountainous territory and too deficient in rainfall for profitable farming. Its economy depended on the collection of customs dues from the caravan trade between Egypt and the Levant.

Edom's most famous, or notorious, son was King Herod the Great (73/74–4 BC) who was created King of Israel by the Romans. He gets an extremely adverse write-up in Jewish sources and is called a mad man who murdered members of his own family, although that was a common occurrence in the ambitious extended families of ancient Near Eastern rulers and nothing to be surprised at. He is blamed for having ordered the Massacre of the Innocents (Matthew 2:16), the killing of all the children of Bethlehem of two years of age and under, although that probably never happened. There were two reasons why Herod was hated by the Pharisaic authority in Jerusalem. For a start he was a client king, imposed on the Judeans by the occupying Roman power. Secondly he was an Edomite, not one of the Children of Israel. The Edomites had adopted the Jewish religion. But the Pharisees of Jerusalem did not recognize them as being Jews. There exists a slightly tongue in cheek on-line King Herod Appreciation Society.

Elamites

Their capital was Susa (in modern Khuzestan) east of the Tigris and Euphrates.

They spoke an agglutinative language that first appeared in 2600 BC. It was not related to any other known language and was written from 2500 BC in a modified form of cuneiform. So it can be read phonetically but not be understood.

Hebrews

When they occupied Canaan in the late Bronze Age they are known as the Israelites, usually distinguished for clarity as citizens of a particular Israelite kingdom. It is an archaeological convention that the Children of Israel who left Egypt in the Exodus are called the Hebrews until they enter the Sinai desert, after which they are called the Israelites. The term Jews is used to refer to them in contexts where their religion is under discussion. Care should be taken to avoid confusion between the people of the united state of Israel under the Judges before the establishment of the United Kingdom of Israel under the kings, Saul, David and Solomon. Their citizens were called Israelites. The citizens of the later northern kingdom of Israel of the Divided Monarchy were also called Israelites in Old Testament times. The citizens of the southern Kingdom of Judah were also Israelites in a general cultural sense, while politically they were Judeans. The term Israeli is a modern 20th century one and is not used of ancient people.

Hittites

They were originally a collection of peoples who spoke three different Indo-European languages, Luwian, Nesite and Palaic. Luwian was written in hieroglyphs, which incidentally is a form of writing that has nothing to do with Egyptian hieroglyphs. The others were written in cuneiform. They occupied and built up a big civilization in central Anatolia between 1650 and 1200 BC in an area in the centre of that land mass. They called it the Land of Hatti. Their capital was the city of Hattusas whose massive remains are now the archaeological site of Boğasköy in central Turkey.

They had a highly efficient army with a large chariot arm, which they used to defend their trading connections

Hurrians

The Hurrians were a prominent and wide-spread people who spoke a language of the Hurrian-Urartian family, which was not related to any other language outside that family. In the 2nd millennium BC they inhabited an area that extended from the north of Assyria, which they controlled briefly, westwards across the northern border of the Syrian desert. Their most important centre of power was the kingdom of Mitanni, which succumbed to Hittite and Assyrian attacks in the 14th century BC after being a strong regional power for two hundred years.

Hyksos

In the Second Intermediate Period of Egyptian history from about 1700 BC, Egypt was invaded and thereafter ruled for around a hundred and fifty years by a grouping of foreign peoples of Anatolian or Palestinian origin generally known as the Hyksos. They introduced chariots drawn by domesticated horses, which had been unknown in Egypt prior to that. They were expelled about 1565 BC by the Egyptian leader Amosis who became the first pharaoh of the New Kingdom. That was the beginning of the Eighteenth Dynasty, and the New Kingdom also included the Nineteenth and Twentieth Dynasties and lasted until 1085 BC.

Israel (not the modern State of Israel)

In 841 BC the Israelites in the northern Jewish state of Israel and the people in the southern state of Judah did not like each other. The origins of the difference between them is buried deep in late Bronze Age tribal politics when they were members of different tribes of the coalition of Israelite tribes who had left Egypt hundreds of years before.

Israel (the State of Israel)

The modern State of Israel did not exist in the Bronze Age. But as it is so much in the news and shares the name Israel with Two Bronze Age kingdoms it would be helpful to present a potted version of its antecedents, leaving out the complexities and being as non-partisan as possible.

After their conquest by the Assyrians, the Kingdoms of Israel and Judah became part successively of the Neo-Babylonian Empire, the Persian Empire, the dominions of the successors of Alexander the Great, and the Roman Empire. The Romans installed Herod the Great as a puppet king.

After the dispersal of the Jews by the Romans who found them ungovernable and the later collapse of the Roman Empire, Palestine became successively part of the dominions of every transient ruler who set out to be the head of a world power. Then in the late Middle Ages it became part of the Turkish Ottoman Empire. After it was on the losing side in the First World War the Ottoman Empire, except for Turkey itself, was divided up among the victors. France got Syria and carved the new republic of the Lebanon out of part of it. Britain got Palestine, Transjordania and Iraq. These were League of Nations Mandates. The Mandatory Powers agreed to lead their mandated countries to independence at some unspecified date. Palestine had a largely Arab population with a Jewish element. Unfortunately, during the First World War the British, in their desperate efforts to find friends who would help them defeat the Turks, had promised Palestine after the War to both the Jews and the Arabs. In 1922 Palestine became a League of Nations Mandated Territory, to be governed by the British. The British found it as difficult to govern as the Romans had. The Jews and the Arabs fought each other and they both fought the British. Increased Jewish immigration into Palestine after the Second World War further complicated the situation and in 1947, wearied of ruling this ungovernable country, the British handed the mandate over to the United Nations. Overnight the Zionists, enthusiasts for a Jewish homeland in Palestine, declared the foundation of the State of Israel. It survives to this day as the United States of America's principal ally in the Middle East.

Jebusites

The Jebusites were a Canaanite people who, according to the biblical narrative, conquered and inhabited Jerusalem before its conquest by King David, possibly in the 11th century BC. The chronology of the period is complicated and much of it is in dispute.

Judea

Judah, or Judea, was a Jewish state with its capital at Jerusalem on the west bank of the Dead Sea separated from the Mediterranean Sea by Philistia. It had kings, as mentioned in the section above on the Hebrews; Saul, David, and then Solomon between the 11th and the 10th centuries BC. After the northern Jewish state of Israel was conquered by the Assyrians in 720 BC Judah remained nominally independent of Assyria, to whom it still had to pay tribute.

The people of the southern Jewish kingdom of Judah did not like the people of the northern Jewish state of Israel whose principal city was Samaria, and considered that they had given in to the Assyrians too easily. The Judean prophets criticized the people of Samaria for every possible form of back sliding from proper Jewish practices.

The point of the New Testament parable of the Good Samaritan, which would have been obvious to any Judean, was that by the Judean way of it, no inhabitant of Samaria could possibly be capable of any generous action.

Kassites

They were a people from originally the Zagros Mountains east of Babylonia who invaded Babylonia after the Hittites sacked the city in 1585 BC and who then ruled Babylonia for the next 400 years. Their linguistic origins are unknown as no texts exist of their language. All we have are word lists that have names of horses and terms for the parts of chariots. Several of their rulers have Indo-European names, so like the Mitannians they might have had an Indo-European aristocracy. Under them, southern Mesopotamia became a state rather than the collection of city states it had been under the Sumerians. They were finally defeated by the Elamites about 1155 BC.

Lydia

This was a kingdom to the west of Phrygia in western Anatolia, which appears briefly in the history of Assyria in reports of the wars the Assyrians had with the Cimmerians.

Medes

They were an Iranian people. They arrived in the area east of Assyria at the end of the 2nd millennium BC. An alliance between them and the

Babylonians and the Scythians enabled them to capture Nineveh in 612 BC. They were conquered in 560 BC by Cyrus the Great who established the Persian Achaemenid Empire.

Mesopotamia

The name is derived from the Greek and means 'Between the Rivers'.

It is the name still used by archaeologists when dealing with the history of what is now the Republic of Iraq. Iraq is a recent country invented in the 1920s by Gertrude Bell and Winston Churchill out of three provinces of the defeated Turkish Ottoman Empire in the carve up of the Middle East between the victorious powers after the First World War.

Minoan civilization

The Minoan civilization was a Bronze Age culture that flourished in Crete between approximately 2500 and 1450 BC. It was named Minoan after the legendary King Minos by the archaeologist Sir Arthur Evans who excavated the palace in the capital, Cnossos between 1900 and 1935 CE. The Minoan culture blended aspects of Egyptian and Asiatic cultures.

Mitanni

Mitanni with its capital, Washukkanni, on the Habur river was the largest and most important Hurrian kingdom. It was founded in the 16th century BC. The Hurrians, speaking a language that has no other known relations except possibly the Urartian of the 1st millennium BC in eastern Anatolia, formed the ordinary people of the kingdom of Mitanni and they were ruled by an Indo-European speaking aristocracy, which was famous for its members' prowess as chariot warriors.

In the early times of Amenophis I and Tuthmosis I of Egypt in the 16th century BC Mitanni was not strong enough to threaten Egyptian interests. But by the time of Tuthmosis III who ascended the throne in 1504 BC it had become sufficiently powerful for Egypt to treat it seriously. Its power lasted for about two hundred years until it was destroyed in the 14th century BC by the Assyrians.

Moab

In Old Testament times the Kingdom of Moab was situated in the Jordanian highlands on the eastern shore of the Dead Sea. The Moabites spoke a dialect of the now extinct Canaanite language and they were often in conflict with the Kingdom of Judah.

Mycenaeans

The Mycenaeans spoke a language that was an early form of Greek, wrote it in a script nowadays called Linear-B and occupied Greece in the 2nd millennium BC. In effect they were Greeks but we call them Mycenaeans to distinguish them from the later Classical Greeks and the even later Macedonian Greeks of the time of Alexander the Great.

Neo-Hittites, also known as Syro-Hittites

These people were the remains of the Hittites who settled in small city states like Carchemish, Aleppo, and Hama in northern Palestine after the collapse of the Hittite Empire in the 12th century BC. Uriah the Hittite in the Bible, after whose wife King David lusted, was one of them. They retained a lot of Hittite culture mixed with the local Canaanite one and spoke Luwian, an Indo-European language of the Hittites as well as Aramaic and Phoenician.

Palestine

Palestine is the common name, slightly out of date now, for the land on the eastern shore of the Mediterranean Sea and between it and the River Jordan. Between 1922 and 1947 it was a British Mandated Territory of the League of Nations and from 1948 to date the State of Israel has been the principal country there.

Philistines

These are a people of whose origin there is no certainty. They are often regarded as being the same as the Peleset, one of the components of the group known as the 'Sea Peoples', immigrants from the Aegean, northern

Levant and Anatolia, who invaded Egypt in the 12th century. When the 'Sea Peoples' were expelled from Egypt by Ramesses III it is possible that the Peleset retreated to the southern part of Canaan where they settled, principally in and around the five cities of Gaza, Ashdod, Askelon, Ekron and Gath. They became known to the Hebrews, with whom they were often at war, as the Philistines and gave the coastal region of the Levant its name Palestine, the land of the Philistines. They appear to have spoken their own original language and also a dialect of Canaanite and to have worshipped Semitic gods such as Dagan, 'Ashteroth and Baal-zebub.

Phrygia

Phrygia, known to the Assyrians as Mushku, was a kingdom that occupied an area east of Hatti in the 11th century BC at the beginning of the Iron Age. We know from the records of the Assyrian king Tiglath-pileser I (1115–1077) that shortly after his accession twenty thousand Phrygians pushed southwards and occupied the Assyrian province of Kummuh. For the next three hundred years the Assyrians fought successive campaigns to limit the extent of this increasingly wealthy kingdom. Its king, Mita, known to the later Greeks as Midas, was renowned for his wealth.

Scythians

They were a semi-nomadic horse-breeding people in the Caucasus who in the 2nd millennium BC moved into the hill country east of Assyria. Esarhadden, King of Assyria, made a temporary peace with the Scythians and married off his daughter to the Scythian chief Bartateatea. Pharaoh Psammetichus I (663–609) avoided Egypt being invaded by the Scythians by paying them tribute. They conquered the Elamites right at the end of the Assyrian Empire, replacing one threat to the peoples of the Tigris-Euphrates plain with another.

Sea People

They were part of the mass movement of people in south-eastern Europe in the 13th century BC. Those whom the Egyptians called 'The People of the

Sea' fled from other expanding people behind them along the coast lines of Asia Minor and Syria and eventually ended up in Egypt. After they were defeated by Ramesses III in 1164 BC some of them entered Egyptian service while others, a group known as the Peleset, settled in the southern part of the Canaanite coast and probably became known as the Philistines and gave the whole area its name, Palestine.

Sumerians

The Sumerians were a people who settled in the southern part of Mesopotamia in small city states around the end of the 3rd millennium BC. Their language has no known relatives. They produced the first writing in cuneiform, punched into soft clay tablets, around 3000 BC. The writing started as stores lists and developed into literature, including an epic of the Flood that was translated into Babylonian and formed the basis of the story of the Flood in the Hebrew Old Testament. The Hebrew version follows the Sumerian original closely. The Sumerians have also left us with the first evidence for the wheel in the Middle East with their two and four-wheeled battle carts and wagons.

Ugarit

The Syrian town of Ugarit, from whose name we have formed the adjective Ugaritic, is situated on the Syrian coast, west-south-west of Aleppo and opposite the island of Cyprus. It was of no great political importance in its day, but it is of interest to the historian because of the number of tablets that were excavated there, which provide us with useful information on Egyptian-Mitannian relations.

Urartu

In the 1st millennium BC Urartu was a thriving state round Lake Van in eastern Anatolia. The language of the people was Urartian, also called Vannic, and in older archaeological publications Chaldean. It was written in cuneiform and belonged to a language family called Asianic, which only means that its speakers came originally from Asia. Urartu is mentioned in history first as Uruatri in the inscriptions of Shalmaneser I (1274–1245) when it was a small part of a loose confederation of states. It became powerful in the 9th century and was a military target of the Assyrian Shalmaneser III (858–824) in the 9th century BC. Attacked by the Cimmerians from the north and Assyria from the south, Urartu was weakened and succumbed finally in the early 6th century BC to attacks by the Scythians and their allies the Medes.

Bibliography

Albright, W. F.: The Date of the Kapara Period at Gozan (Tell Halaf). *Anatolian Studies* 6, 1958. pp. 75–85.

Anthony, David W.: *The Horse the Wheel and Language*. Princeton University Press, Princeton N. J. 2007.

Anthony, David W. and Brown, Dorcas R.: Eneolithic Horse Exploitation in the Eurasian Steppes, Diet, Ritual and Riding. *Antiquity* 74, 2006. pp. 75–86.

Anthony, David and Vinogradov, Nikolai: Birth of the Chariot. *Archaeology*. 48, 2, 1995. pp. 36–41.

Barnett, R. D.: *The Assyrian Palaces and their Sculptures in the British Museum*. Batchworth Press, London. (n. d.).

Barnett, R. D. and Falkner, M: *The Sculptures of AŠŠUR-NAṢIR-APLI II (883–859 BC) TIGLATH-PILESER III (745–727 BC) ESARHADDON (681–669 BC) from the Central and South-West Palaces at Nimrud*. British Museum, London. 1962.

Bökönyi, Sándor: *The Przevalsky Horse*. Souvenir Press, London. 1974.

Bökönyi. Sándor: The Earliest Waves of Domestic Horses in East Europe. *Journal of Indo-European Studies* 6, 1978. p. 17.

Bryce, Trevor: *Hittite Warrior*. Osprey, Botley, Oxford. 2007.

Buttery, Alan: *The Armies and Enemies of Ancient Egypt and Assyria*. Wargames Research Group. 1974.

Cazelles, H.: The Hebrews. In Wiseman, D. J. (ed.). *Peoples of Old Testament Times*. Oxford University Press, Oxford. 1973. pp. 1–28.

Chenevix Trench, Charles: *A History of Horsemanship*. Longman, London. 1970.

Clausewitz, Major General Carl von: *On War*. Oxford World Classics, Oxford. 2008.

Crouwel, J.H.: Aegean Bronze Age Chariots and their Near Eastern Background. *Bulletin of the Institute of Classical Studies*. 25, 1978. pp. 174–175.

Cruden, Alexander: *Cruden's Complete Concordance to the Old and New Testaments*. Henrickson Publishers Marketing. Peabody, Massachusetts, 2012.

Dalal, Anita: *Ancient India*. National Geographic Society, Washington DC, 2007.

Drews, Robert: *The End of the Bronze Age. Changes in Warfare and the Catastrophe. CA 1200 BC*. Princeton University Press, Princeton N.J. 1993.

Drower, M. S.: The Domestication of the Horse. In Peter J. Ucko and G. W. Dimbleby (eds): *The Domestication and Exploitation of Plants and Animals*. Duckworth, London. 1969. pp. 471–478.

Fields, Nic: *Bronze Age War Chariots*. Osprey, Botley. Oxford. 2006.

Grossman, Lieutenant Colonel Dave: *On Killing. The Psychological Cost of Learning to Kill in War and Society*. Back Bay Books, New York. 1996.

Groves, Colin P.: *Horses, Asses and Zebras in the Wild*. David & Charles, Newton Abbot. 1974.

Gurney, O. R.: *The Hittites*. Penguin Books, Harmondsworth. 1962.

Hackett, General Sir John (ed.): *Warfare in the Ancient World*. Sidgwick & Jackson, London. 1989.

Hamblin, William J.: *Warfare in the Ancient Near East to 1600 BC Holy Warriors at the Dawn of History*. Routledge, London and New York. 2011.

Hawkins, J. D.: Assyrians and Hittites. *Iraq*. Vol 36, 1974. pp 67–84.

Healy, Mark: *New Kingdom Egypt*. Osprey, Botley, Oxford. 1982.

Healy, Mark. *The Ancient Assyrians*. Osprey, Botley, Oxford. 1991.

Hoffner, H. A.: The Hittites and Hurrians. In Wiseman. *op. cit*. pp. 197–228.

Jankovich, Miklós: *They Rode into Europe*. Harrap, London. 1971.

Karasulas, Antony: *Mounted Archers of the Steppe 600 BC–AD 1300*. Osprey Publishing, Botley, Oxford. 2004.

Kitchen, K. A.: The Philistines. In Wiseman. *op. cit*. pp. 53–78.

Kuznatsov, P. F.: The Emergence of Bronze Age Chariots in Eastern Europe. *Antiquity* Vol 80, No. 309. 2006. pp. 638–645.

Labat, René: *Manuel d'Epigraphie Accadienne*. Imprimerie Nationale, Paris, 1963.

Lacheman, Ernest R.: Epigraphic Evidence of the Material Culture of the Nuzians. In Starr *op. cit*. Appendix D.

Lambert, W. G.: The Babylonians and Chaldeans. In Wiseman. *op. cit*.

Levine, Marsha A.: Dereivka and the Problem of Horse Domestication. *Antiquity* 64, 1990.

Liancheng, Lu: Chariot and Horse Burials in Ancient China. *Antiquity* 67, 1993. pp. 824–838.

Littauer, M. A: Bits and Pieces. *Antiquity* XLIII. 1969. pp. 289–300.

Littauer M. A. and Crouwel J. H.: *Wheeled Vehicles and Ridden Animals in the Ancient Near East*. Brill, Leiden. 1979.

Lloyd, Seton: *Foundations in the Dust. A Story of Mesopotamian Exploration*. Penguin Books, Harmondsworth. 1955.

Lloyd, Seton: *Early Highland Peoples of Anatolia*. Thames and Hudson, London. 1967.

Luckenbill, D. D.: *The Annals of Sennacherib*. Oriental Institute of the University of Chicago, Chicago Ill. 1924.

Luckenbill, D. D.: *Ancient Records of Assyria and Babylonia*. Vol I. Historical Records of Assyria from the Earliest Times to Sargon. Oriental Institute of the University of Chicago, Chicago Ill. 1926.

Luckenbill, D. D.: *Ancient Records of Assyria and Babylonia*. Vol II. Historical Records of Assyria from Sargon to the End. Oriental Institute of the University of Chicago, Chicago Ill. 1927.

Macalister, R. A. Stewart: *The Philistines. Their History and Civilisation*. Oxford University Press, Oxford. 1914.
Madhloom, T. A.: *The Chronology of Neo-Assyrian Art*. Athlone Press, London. 1970.
Mallowan, M.E.L.: *Early Mesopotamia and Iran*. Thames and Hudson, London. 1965.
Millard, A. R.: The Canaanites. In Wiseman, *op. cit.* pp. 29–52.
Moorey, P.R.S.: The Emergence of the Light, Horse Drawn Chariot in the Near East 2000 – 1500 BC. *World Archaeology* 18, 1986. pp.196 – 215.
Moorey, P, R. S.: Pictorial Evidence for the history of horse riding in Iraq before the Kassite Period. *Iraq*. XXXII, 2014, pp. 36–50.
Noble, Duncan: *The Development of Transport in Ancient Mesopotamia* (unpublished PhD thesis, University of London, 1976). London. 1976.
Noble, Duncan: The Mesopotamian Onager as a Draught Animal. In Peter J. Ucko and G.W. Dimbleby (eds) *The Domestication and Exploitation of Plants and Animals*. Duckworth, London. 1969. pp. 485–488.
Noble, Duncan: Assyrian Chariotry and Cavalry. State Archives of Assyria Bulletin of the University of Padua, Padua. Vol IV. Issue 1. 1990.
Piggott, Stuart.: The Earliest Wheeled Vehicles and the Caucasian Evidence. *Proceedings of the Prehistoric Society* 34, 1968. pp. 259–266.
Piggott, Stuart: Chariots in the Caucasus and China. *Antiquity* Vol 48, No. 189, March 1974. pp. 16–24.
Piggott, Stuart: Bronze Age Chariots in the Urals. *Antiquity* Vol 49, No. 196, December 1975. pp. 289–90.
Piggott, Stuart: *The Earliest Wheeled Transport from the Atlantic Coast to the Caspian Sea*. Thames and Hudson, London. 1983.
Ramkumar, N: *Harappa Civilization: Who are the Authors?* Privately published, 2012.
Rudenko, Sergei I.: *Frozen Tombs of Siberia. The Pazyryk Burials of Iron Age Horsemen*. Dent, London. 1970.
Roux, Georges: *Ancient Iraq*. Pelican Books, London. 1992
Saggs, H. W. F.: *The Greatness that was Babylon*. Sidgwick and Jackson, London. 1962.
Saggs, H. W. F. The Assyrians. In Wiseman, *op. cit.* pp.156–178.
Salonen, Armas: *Hippologica Accadica. Suomalaisen Tiedeakatemia Toimituksia* (Annals of the Finnish Academy of Sciences). Helsinki. 1951.
Salonen, Armas: *Die Landfahrzeuge des Alten Mesopotamien.* Suomalaisen Tiedeakatemian Toimituksia (Annals of the Finnish Academy of Sciences). Helsinki. 1956.
Sawyer, Ralph D. (tr), (ed): *The Essence of War. Leadership and Strategy from the Chinese Military Classics*. Westview Press, Boulder, Colorado. 2004.
Sawyer, Ralph D.: *Ancient Chinese Warfare*. Basic Books, New York. 2011.

Selby, Stephen: *Chinese Archery*. Hong Kong University Press, Hong Kong. 2013.

Shaughnessy, Edward L.: Historical Perspectives on the Introduction of the Chariot into China. *Harvard Journal of Asiatic Studies*. Harvard, Conn. Vol 48.1, 1988. pp. 139–237.

Sidnell, Philip: *Warhorse*. Hambledon Continuum, London. 2006.

Starr, Richard F.S.: *Nuzi. Report on the Excavations at Yorgan Tepa, Near Kirkuk, Iraq*. (2 vols). Harvard University Press, Harvard, Conn, 1939.

Stillman, Nigel and Tallis, Nigel: *Armies of the Ancient Near East 3,000 BC to 539 BC*. Wargames Research Group. 1984

Sun Tzu: *The Art of War*. (tr) Lionel Giles 1910. Simon & Brown, Hollywood, Florida. 2010.

Watkins, Trevor: The Beginnings of Warfare. In Hackett, *op. cit.* pp. 36–53.

Wheeler, Sir Mortimer: *Civilizations of the Indus Valley and Beyond*. Thames and Hudson, London. 1966.

Wise, Terence: *Ancient Armies of the Middle East*. Osprey, Botley, Oxford. 1981.

Wiseman D. J.: The Assyrians. In Hackett, *op. cit.* pp. 36–53.

Wiseman, D. J. (ed.): *Peoples of Old Testament Times*. Oxford University Press, Oxford. 1973.

Woolley, Sir. Leonard: *Ur; The First Phases*. King Penguin Books, London. 1946

Yadin, Yigael: *The Art of Warfare in Biblical Lands*. Weidenfeld and Nicolson, London. 1963.

Zeuner, F.: *A History of Domesticated Animals*. Hutchinson, London. 1963.

Index

Modern geographical names are in italics